# The New View of the Earth

## Moving Continents and Moving Oceans

**Seiya Uyeda**

University of Tokyo

*Translated by Masako Ohnuki*

W. H. Freeman and Company
San Francisco

新しい地球観：動く大陸・動く大洋

*Translator's Note:* I acknowledge with gratitude the generous help of Dr. Hiroo Kanamori and Ms. Lilla Weinberger, without whom my translation could not have been completed. —*M. O.*

**Library of Congress Cataloging in Publication Data**

Uyeda, Seiya, 1929–
    A new view of the Earth.

      Translation of Atarashii Chikyukan.
      Bibliography: p.
      Includes index.
      1. Continental drift. I. Title.
QE511.5.U93-1-3     551.1'3     77-9900
ISBN 0-7167-0283-5
ISBN 0-7167-0282-7 pbk.

ATARASHII CHIKYUKAN (The New View of the Earth)
              by Seiya Uyeda

© 1971 by Seiya Uyeda
Originally published in Japanese by IWANAMI SHOTEN, Publishers, Tokyo.

1  2  3  4  5  6  7  8  9

# Preface

This book is a translation of the original Japanese edition *Atarashii Chikyukan (The New View of the Earth)*, published in 1972. For the English version, I have made extensive additions, in both the text and the illustrations, to incorporate still more recent developments in our current understanding of the earth, and I have omitted some material that did not seem relevant to readers outside of Japan.

A number of my friends contributed to the initial writing of the book: in particular, Dr. Hiroo Kanamori, Dr. Kazuaki Nakamura, Dr. Masashi Yasui, and my wife Matsuko Uyeda all had valuable comments to make on the Japanese edition, and Sir Edward Bullard and Dr. Frank Press had additional suggestions for this one. Dr. Allan Cox has gone through the text with meticulous care, and greatly improved both its science and its language. Robert Geller and Seth Stein read the proofs and suggested a number of important changes that I have incorporated into the final book. I also thank my translator Mrs. Masako Ohnuki and my editor Michele Liapes for their painstaking efforts.

*September, 1977*                                        Seiya Uyeda

# Contents

**Preface**   iii

**Introduction**   1

**Chapter 1**   The Theory of Continental Drift: Its Birth, Death, and Revival   6

**Chapter 2**   The Exploration of the Ocean Floor   42

**Chapter 3**   The Hypothesis of the Spreading Ocean Floor: A Synthesis   62

**Chapter 4**   Plate Tectonics   93

**Chapter 5**   Island Arcs   125

**Chapter 6**   The New View of the Earth   172

**Typical Rocks of the Continental Crust, Oceanic Crust, and Upper Mantle**   204

**Bibliography**   206

**Index**   211

# Introduction

序

## Rapid Progress Versus Eternal Questions

In recent years truly dramatic discoveries in scientific research have
unfolded, and progress has accelerated at a stunning pace. One reason is the advent of the electronic computer, which enables us to
compile enormous amounts of data into meaningful categories, so
that the multitudes of observations can be applied to the development
of significant and universal concepts. The current abundance of new
ideas is evident in the succession of papers now being written. Of the
modern sciences, the field of earth science is beginning to be the
most rapidly advancing. In particular, that area concerned with the
solid part of the earth has recently undergone phenomenal changes of
the sort that occur only rarely in any one field. They are remarkable
changes—of interest to the layman as well as to the scientist—and I
shall describe them in the chapters that follow.

The rapid progress does not mean, however, that all the questions
have been completely resolved. In fact, recently, while still writing
this book, I recalled an article of the late 1940s enumerating six
important unresolved questions that were important at the time.
It was not until I contacted my old friend Keiiti Aki of the
Massachusetts Institute of Technology, with whom I had read and
discussed the article so many years ago when both of us as
undergraduates were looking for problems to tackle, that I was
able to locate it. He promptly forwarded a copy to me, thanking me
for reminding him of the memorable piece. It was a lecture entitled
"Some Unsolved Problems of Geophysics" by L. H. Adams (1947)
the president of the American Geophysical Union, and the six problems were the following:

(1) the origin of the mountain chains;
(2) the origin of geosynclines (deep basins filled with sediments);
(3) the cause of volcanoes and other igneous activity;
(4) the cause of deep earthquakes;
(5) the origin of the earth's magnetic field;
(6) the temperatures prevailing in the interior of the earth.

Although these problems were not the *only* existing questions of importance, all six were indeed significant. Furthermore, none of them have yet been completely solved; all of them remain as important as ever. To be sure, some people may argue, "Look at how many of these problems we *have* solved!" But have we really? This issue is going to be one of the main themes of this book. Indeed, as any earth scientist will agree, these problems have been considered crucial not only since the 1940s, but for hundreds of years. The progress of earth science may not be as rapid as it seems after all.

## Some Peculiarities of Earth Science

How is it that, despite the rapid progress, we have not answered many fundamental questions about the earth? The fact is that many basic problems of earth science defy answer by means of direct experiments. This poses considerable difficulties for us. Consider, for instance, the sixth of Adams' problems—the temperature of the earth's interior. It is simply not possible—at least at the present time—to *measure* the temperature at the center of the earth. One can merely deduce it indirectly from other evidence.

Then, continental drift and movements within the earth—the central topics of this book—are beset by similar difficulties. The mere penetration of the deep interior of the earth would itself be a difficult task, but the measurement of movements at great depth is beyond the capability of any known instrument or method because of the vast scale of the deep movements and their extreme slowness. Consequently it is very difficult for us to prove by direct observation the actual existence of such phenomena. Even if we can eventually detect the present movements of the continents in relation to one another by precise geodetic measurements of distances, using such instruments as a laser reflector on the surface of the moon, this will in no way prove that similar continental movements occurred in the remote geological past. The continental shifts, splits, and collisions that have occurred throughout the earth's history are once-in-a-lifetime phenomena; and the components involved in these movements have been too massive, and the period of time too vast, for reproduction in the laboratory.

These, then, are the problems that make the basic issues of earth science so difficult to resolve despite the new discoveries that have been made and the storehouse of information that is now accumulating. Indeed, they might lead us to wonder if research in such a field has any validity at all. The potential solution to a problem sometimes seems to become more elusive as research advances, so that the gap between them remains virtually unchanged. However progress *is* being made because our understanding of the scope and significance of the basic problems in earth science is becoming more profound. Perhaps we are not learning to ask better questions, but at least we are beginning to understand more clearly the meanings of our questions.

The difficulty of direct verification is a serious one in earth science, and almost inherent in the field. Yet advances in observation and theory justify more and more the use of indirect verification and bring us increasingly nearer to the truth.

## The Uniqueness of Earth Science

Paradoxically, the fact that solutions to these problems have been hard to come by may have helped earth scientists. We have been obliged to examine and observe patiently many seemingly unrelated phenomena. These observations have led us to propose daring new hypotheses. Then, to prove them, we have had to seek additional and unshakable observations to support them, no matter how indirect such observations might be. The successful combination of the basic field work of earth science with the more abstract concepts of physics and chemistry has been a triumph that geologists savor. The unknown mechanisms—such as those responsible for the origin of the earth, for convection within the mantle, for the origin of the earth's magnetic field, and for deep-focus earthquakes—seem to offer a special challenge and appeal to the earth scientist. The approaches to this challenge vary widely, as do the intellectual tastes and motivations of individuals: for there are many types of people, and human wisdom is advanced by the efforts of all of them.

## Insight

Scientific research includes various types of work. At least two processes are necessary—(1) the accumulation of data by experiment and observation, and (2) the analysis and theorization of that data. Many individual researchers tend to focus on one or the other of these processes, depending on their own inclinations. The modern age demands highly refined skills in any one aspect of the work, thereby creating niches for people who specialize in one of the following categories: experimentation, observation, analysis, or theory. Such specialization is unavoidable to a certain extent. A true researcher, however, should not allow himself to become immersed in one to the exclusion of the others. For example, it is possible for one to get so absorbed in taking measurements day and night, that he forgets to think. But if his work is to be really valuable, he must back his efforts by sound reasoning.

Indeed, good research requires a deep understanding of the scope of basic problems and a high degree of trained perception. Only too often superficial ideas and notions are mistaken for genuine creative thought. Although such ideas in themselves should not be discouraged, what the theoretician really needs is the special ability called *insight*—that capacity to select the genuinely promising idea from the others and to develop it into a theory or a set of predictions that can be experimentally verified. It is the most important quality a scientist can have.

The concepts treated in this book are the results of true scientific insight. It was insight that produced an important shift in our perspective of the earth—from a *fixist* view of an unchanging and stable body to a *mobilist* view of drifting land masses and ocean basins.

**Outline of the Book**

Throughout the book, general background in earth science will be provided for lay readers. In the first chapter we will discuss the history of the theory of continental drift, first introduced by Alfred Wegener. It was a theory that enjoyed temporary popularity after its introduction, followed by denunciation and rejection as an almost heretical idea. It then experienced a dramatic revival after World War II, owing to the introduction of paleomagnetism, which is the study of the history of the earth's magnetic field by means of the natural magnetization of rocks.

In the second chapter we will outline the findings in ocean floor geology that helped to revive the theory of continental drift. In this field, too, fantastic progress has been made since World War II, yielding an enormous amount of information that we could not have obtained from research limited to the land.

The theories of sea-floor spreading and plate tectonics will be outlined in the third and fourth chapters. The theory of sea-floor spreading is based on the idea that the ocean floor is created at the mid-oceanic ridges, spreads out horizontally, and disappears in the deep trenches. This theory was a remarkable synthesis of numerous independent data, and its dramatic success resulted in the even more fascinating concept of plate tectonics. Basically, the earth's surface is thought to consist of about 10 plate-like solid blocks, approximately 70 kilometers thick, which interact with one another. A further application of this concept to the past suggests that such interactions have been the primary cause of mountain building and other large-scale movements of the earth's crust throughout the earth's geologic history.

In Chapter 5 we shall examine island arcs from the new perspective of plate tectonics, using Japan as an example. The Japanese island arcs form part of a circum-Pacific belt of volcanoes, large earthquakes, deep trenches, and faults that are assumed to have been caused by the underthrusting of the floor of the Pacific Ocean beneath the Asian continent.

In the last chapter, we will describe the transition from the so-called fixist to the mobilist view of the earth, along with the possible driving mechanisms of plate tectonics.

# Chapter 1

# The Theory of Continental Drift: Its Birth, Death, and Revival

**Wegener's Idea**

In 1912 the German scientist Alfred Wegener (1880–1930) proposed a new theory.* He maintained that the continents on either side of the Atlantic—the North American and South American continents and the European-African continent—were once joined and that they had split and drifted apart into their present positions. He insisted that all other continents, including India, Australia, Africa, and Antarctica, also belonged to the one gigantic protocontinent. He named this great hypothetical continent *Pangaea*. Figure 1-1 illustrates the process of continental breakup. Wegener believed that Pangaea was united until the late Carboniferous period, about 300 million years ago, and then began to split apart, ending up in the present distribution of continents. Since Pangaea was the only continent, it was surrounded by one enormous ocean. No individual oceans, such as the Atlantic, Indian, or Antarctic, existed at that time. This was the essential idea of continental drift. It was a spark that generated a new view of the earth.

The source of Wegener's idea was the realization that the outlines of the continents fit like the pieces of a jigsaw puzzle. This conformity can be seen by anyone who looks closely at the coastlines along

---

*It is true that the idea of continental drift dates back, long before Wegener, to A. Snider (in 1858) and even to F. Bacon (in 1620). But it was Wegener who first made the case an important scientific issue.

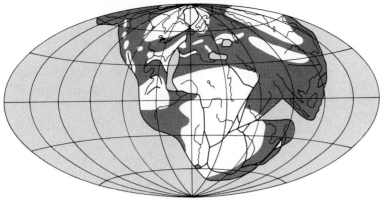

Late Carboniferous (300 million years ago)

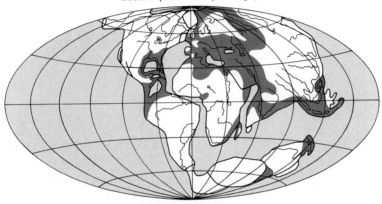

Eocene (50 million years ago)

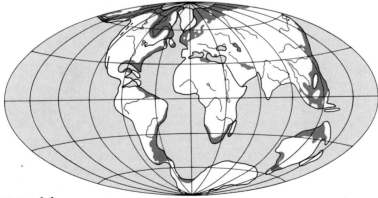

Early Pleistocene (1.5 million years ago)

**FIGURE 1-1**
Reconstruction of the map of the world for three periods according to Wegener's theory of continental drift. Africa is placed in its present-day position as a standard of reference. The heavily shaded areas represent shallow seas. Ages in millions of years have been added. [After A. Wegener, *The Origin of Continents and Oceans.* Methuen, London, 1924.]

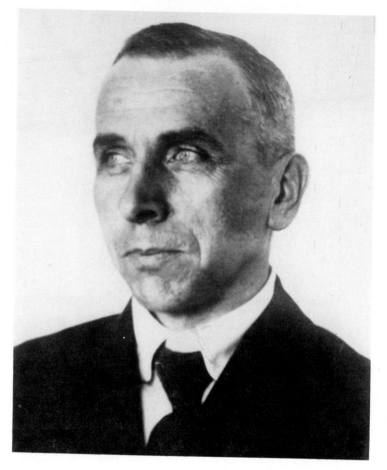

FIGURE 1-2
Alfred Wegener. [Photo from Historical Pictures Services, Inc., Chicago.]

the Atlantic Ocean. The idea, simple as it was, was considered pre-posterous at the time because it conflicted with the universal belief that the earth was immobile.

Wegener (Figure 1-2), a meteorologist by profession, was one of the pioneers in the field of high-altitude meteorological observation. Also his exploration of the previously unpenetrated continent of Greenland contributed to the research in this area. The culmination of these diverse activities was the conception and development of his theory of continental drift. It began as a simple idea, but Wegener did not allow it to remain as such: he pursued it resolutely and systematized the theory. It was because of this perseverance that he was a great scientist. Ideas occur to any scientist from time to time. But he fails to develop most of them, and then forgets them because they seem either too fantastic or impractical. The majority of them are indeed useless. Wegener confessed that he himself considered the possiblity of continental drift to be fantastic and impractical, and at first did nothing about it. However, unlike many scientists who abandon interesting ideas and regret it later, Wegener began to develop his seemingly simple theory. His search for new knowledge started with the study of geology and paleontology, fields remote from his speciality. This project, conceived in 1910, was interrupted by his expeditions to Greenland and his military service during World War I, in which he was injured. Yet such obstacles did not deter him. In 1915 he published his monumental work, *Die Entstehung der Kontinente und Ozeane (The Origin of Continents and Oceans)*, and by 1923 had revised it three times. In 1924, he published *Die Klimate der Geologischen Vorzeit (The Climate Through Geological Time)* with a meteorologist, W. Köppen. During this period he also published a great many other papers. These works were the fruit of his revolutionary view of the earth as developed from the concept of continental drift. It was as if modern solid earth science had evolved within the mind of this one man who was tens of years ahead of everyone else.

## The Geologic Method

As a meteorologist, Wegener needed more than anything else a knowledge of geology for his pursuit of the history of continental drift. We, too, need an understanding of the basics of geology in order to grasp Wegener's ideas.

The two following fundamental principles that geologists apply when studying the history of the earth are particularly important:

(1) *The law of superposition.* If one stratum (or layer) overlies another, the top stratum is younger than the bottom one.

(2) *The law of faunal assemblage.* Strata that contain fossils of the same species of animals and plants were produced in the same period.

The first law is self-evident: without the existence of the prior stratum, the new stratum could not be deposited on top of it. This law enables us to detect the chronological relationships of the stratified rocks in one place.

The law of faunal assemblage gives us clues about the time relationships among strata scattered in different places. Everyone knows that all forms of life are constantly undergoing evolution. The process might seem slow to us, but considered on a geological time scale it is actually quite rapid. Primitive life forms first appeared on the earth about three billion years ago, and gradually evolved into more complex creatures. This one-way trend of evolution—from the simple to the complex—has enabled us to identify the chronological age of the strata by the fossils (such as trilobites and dinosaurs) preserved within them. Figure 1-3 gives the geological ages as determined by the fossils of animals and plants. The study of fossils is called paleontology and constitutes quite an elaborate system of science. These names of geological eras, periods, and epochs—each with its own legitimate and interesting origin—will be mentioned throughout this book.

Informative though it is, the paleontologic method has two intrinsic limitations. The first is the amount of time that we can go back. As shown in Figure 1-3, it is only in the strata of the past 600 million years or so that plant and animal fossils are complex enough that we can use them to compare the ages of the strata. There are not enough fossils in the older strata to date them. This early period with few or no fossils is in a way a prehistoric, or biological, "dark age," and is called the Precambrian era. The second limitation is that it cannot provide us with "absolute" chronology, since it uses the evolution of animals and plants as its clock. It can determine, for instance, that stratum A is older than B, but it cannot tell us how old either stratum is or how much older A is than B.

Such limitations have been overcome in recent years, thanks to the development of methods of absolute age determination. From the spontaneous disintegration of such radioactive elements as the uranium, thorium, strontium, and potassium that are contained in rocks in small amounts, we can determine the absolute age of the rocks. These radioactive elements constantly and regularly transform into other elements in accord with what is called the law of disintegration. This transformation can be considered a kind of evolution, too, but unlike that of plants and animals, the exact rate of transformation has been determined by physicists. The absolute ages given in Figure 1-3 have been obtained by this method.

Geological strata consist of either igneous rocks or sedimentary rocks. Igneous rocks are primary rocks formed by the cooling and

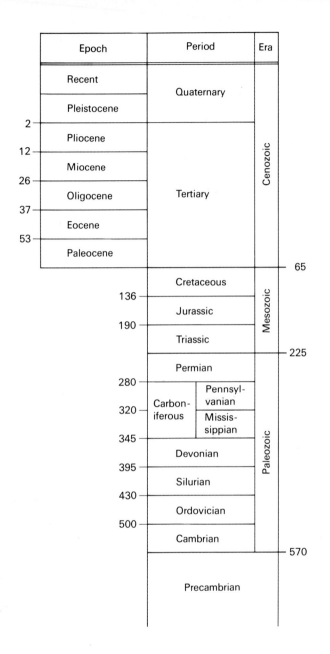

| Epoch | Period | | Era |
|---|---|---|---|
| Recent | Quaternary | | Cenozoic |
| Pleistocene | | | |
| Pliocene | Tertiary | | |
| Miocene | | | |
| Oligocene | | | |
| Eocene | | | |
| Paleocene | | | |
| | Cretaceous | | Mesozoic |
| | Jurassic | | |
| | Triassic | | |
| | Permian | | Paleozoic |
| | Carbon- iferous | Pennsyl- vanian | |
| | | Missis- sippian | |
| | Devonian | | |
| | Silurian | | |
| | Ordovician | | |
| | Cambrian | | |
| | Precambrian | | |

2
12
26
37
53
65
136
190
225
280
320
345
395
430
500
570

FIGURE 1-3
The geologic time scale. The numbers at the sides of the column are ages in millions of years. [After F. Press and R. Siever, *Earth*. W. H. Freeman and Company. Copyright © 1974.]

solidification of magma; sedimentary rocks are secondary rocks formed as a result of erosion and deposition. Sediments are called secondary because most of the particles transported by water and deposited were originally parts of other rocks on land. Most of the rocks we see in strata at the present time are sedimentary rocks. Thus the surface of the land is almost completely covered with sedimentary rocks—even mountain ranges as high as the Alps and the Himalayas, meaning that these great mountain tops were once under water!

Suppose some region is elevated high above sea level: at this point in time deposition ceases and erosion takes over. Even the highest mountains are gradually eroded into level land. The history of an elevated region that has undergone erosion is very difficult to trace because no sedimentary record exists and the history can be studied only indirectly through the record of erosion. If this region is submerged once again, the deposition process will resume and a more complete record of the geologic history will start to accumulate. Any geologist knows for a fact that the rocks forming many of the mountains were once deposited underwater, but the initial idea can be quite a shock. I remember well my own surprise when I first heard it.

## The Land Bridge

If the continents now scattered in the world oceans once formed a single enormous continent, the strata that existed before the breakup would have to be related to one another. Moreover, the strata that formed after the split would be unrelated. To establish this hypothesis and thus confirm his theory of continental drift, Wegener set out to gather evidence. His skepticism about the concept, he explains, was overcome when he came across a paleontological paper discussing the possibility that Brazil had once been linked to Africa. It came as a surprise to Wegener to find that such an assertion had already been put forth, quite independently of his hypothesis of continental drift. But it is exactly this point that I find interesting, because it seems to demonstrate the importance of the *perspective* from which one interprets scientific data. As an amateur in paleontology, Wegener was unaware of any evidence suggesting the ancient connection among continents, and yet paleontologists had long been studying this very possibility. The established interpretation of this concept, however, was entirely different from Wegener's. It was the land-bridge theory.

Having surveyed the distribution of fossils of such animals as monkeys, earthworms, and snails, and of various kinds of plants, paleontologists observed that close affinities prevailed between Africa and South America, Europe and North America, Madagascar and India. For example, since such organisms as snails cannot swim across vast oceans, it was presumed that two continents containing nearly identical snail fossils must have once been connected by land—a land bridge. Whereas Wegener interpreted this distribution as indication that a single continent had once existed and subsequently split into several parts, the traditional paleontological interpretation of the same phenomenon assumed the immovability of continents and thus the existence of a land bridge. The observed phenomena were the same, but they were interpreted from different viewpoints, so that two vastly different theories resulted.

If land bridges had indeed connected the continents, the one joining Africa and South America could not have been a long, narrow protrusion across the Atlantic. Here, one of continental scale would seem more likely. Since this hypothetical bridge no longer existed, the task was to explain the disappearance of such a land mass. The most popular way of accounting for its immersion was to ascribe it to a grandiose depression of the earth's crust. Thus, the land bridge theory assumed that a land mass the size of a continent can "become" a sea. This view was essentially the same as the one that asserted that the distribution of land and ocean is determined by the *vertical* movement of the earth's crust. In the theory of continental drift, the *horizontal* movement of continents is the central phenomenon. This is the fundamental difference in the two hypotheses.

## The Earth's Crust

Before discussing continental drift further, let us examine the nature of the earth's crust. Of the various ways of examining the interior of the earth, the most direct is drilling. Drilling a hole much deeper than 10 kilometers, however, is beyond our present technology. The next most direct method is to survey the earth's interior by studying the propagation of earthquake waves. When an earthquake occurs, seismic waves, originating at the focus of the quake, travel through the earth's interior. Seismic waves are of two basic types. The first type, primary or $P$ waves, travel through the earth just as sound waves travel through the air. $P$ waves transmit the changes in *volume*, the alternating compression and expansion of the earth. Secondary or $S$ waves transmit the distortion of the *shape* of the earth's material. The particle motion of $P$ waves is in the direction of propagation, and that of $S$ waves is perpendicular to the direction of its propagation. The distinction between the two types of waves can be seen in Figure 1-4. $P$ waves travel faster than $S$ waves (approximately 1.7 times faster).

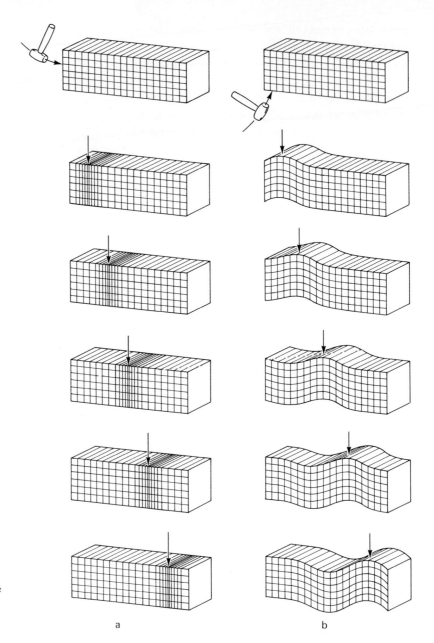

FIGURE 1-4
The passage of a seismic
wave pulse in a block of
material. Repeated hammer
blows or cyclic shock will
generate a train of waves. The
arrow indicates the crest of
the wave: (a) $P$ wave; (b) $S$
wave.

a                          b

The earthquake is usually felt in two successive shocks: first, a light
jolt and then a heavier rocking one. The first indicates the arrival of
the $P$ waves and the second, the $S$ waves. $S$ waves can travel through
solid material but not through fluid, whereas $P$ waves can travel
through both solid and fluid substances.

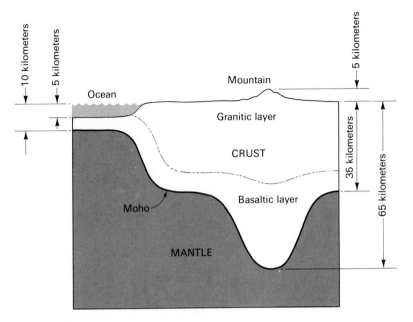

FIGURE 1-5
Schematic cross section of the earth's crust.

Investigation of the propagation of seismic waves originating from either natural or artificial earthquakes reveals that above a certain depth the waves travel slowly. But once the wave reaches this depth, its velocity increases sharply: the velocity of $P$ waves, for example, jumps from about six or seven kilometers per second to about eight kilometers per second.

The surface layer above this boundary is called the *crust*, and the layer below it is called the *mantle*. The boundary itself is called the *Mohorovičič discontinuity* after the Yugoslavian seismologist who discovered it in 1909. This discontinuity is often called simply the Moho or M discontinuity.

The continental crust (from 30 to 50 kilometers thick) is much thicker than that of the ocean floor, which is only several kilometers thick, as shown in Figure 1-5. The crust in continental regions consists of an upper layer of granitic rocks and a lower layer of basaltic rocks in the form of what geologists call *gabbro*. These rocks vary widely in chemical and mineral composition. Also, rocks of similar composition can vary in texture, depending on the mode of formation. All these differences have generated numerous names that are often unfamiliar to laymen. We will be concerned here with only a few of the major rock *types*, which are listed in the table at the end of the book. Rocks of basaltic composition are quite different from granitic rocks. The basalts are dark, fairly heavy, and "primitive" in the sense that they have formed from magma derived directly from the mantle. Granitic rocks are lighter both in color and density, and many have a chemistry suggestive of geologic "recycling." Although much of the crust is too deep to be sampled directly, we have been able to guess at their composition by comparing the seismic velocity (the velocity of earthquake-produced seismic waves traveling through these layers) with seismic velocities in rocks of known composition measured in the laboratory. Therefore, when we say, for example, that the upper layer of continental crust consists of granitic rocks, what we mean is that the layer has the same seismic wave velocity as granitic rocks. It may not really be granitic rock, because rocks with different compositions can have the same velocity. Under the oceans, the crust consists of a thin top layer of sediments and two underlying layers called the second and the third layers. The second layer is presumably composed of volcanic, or *extrusive*, rocks such as basalt, and its *intrusive* equivalent, gabbro.* Existence of basalt at the top part of the second layer has been verified by the Deep Sea Drilling Project (described in Chapter 3). It is not yet known what the third layer consists of. It too may be either a form of gabbro or a rock called serpentinite (see the table). Thus the ocean crust is distinguished by its relative thinness and its lack of a granitic layer. (Apparently rocks that are much heavier than granite—such as peridotite and eclogite—compose the upper part of the mantle.) The thinness of the ocean crust has been verified by seismic experiments only since the 1950s. Consequently the structure of the crust as conceived of in Wegener's day was not exactly the same as that shown in Figure 1-5, owing to the lack of data. However, geologists at that time already had the right idea. They had concluded that the continental crust was substantially different from the crust beneath the ocean floor; they also suspected that there were no continents with thin "oceanic" crusts nor ocean floors with thick "continental" crusts.

This belief was supported by gravity studies, which revealed that underneath a region of elevated topography is a buried root of low-density material. Since the crust consists of rocks lighter than the

---

*Extrusive* rocks are formed when volcanic magma cools at the earth's surface. They can be recognized by their texture, which is either glassy or fine-grained as a result of rapid cooling. *Intrusive* rocks are formed when the magma cools and solidifies at depth, and they can be recognized by a coarser texture consisting of larger grains, a result of slow cooling. *Gabbro* is an intrusive rock type and is the equivalent of volcanic basalt.

material of the mantle, this phenomenon was interpreted as an indication that the crust is thicker where the earth's surface is higher. In a sense, the crust seems to be floating on the mantle, much like an iceberg in the ocean. According to Archimedes' Principle, any iceberg must have a deep root in order to maintain its buoyancy. The loftier the iceberg, the deeper its root. Apparently this principle also applies to the crust: the elevated continents have thicker crusts than the low-lying oceans. This phenomenon is called *isostasy*. It signifies that the mere presence of water does not make an ocean: rather, it is the structural difference in the earth's interior that is responsible for the division of land and ocean. A continent cannot sink to make an ocean, as long as the basic law of buoyancy holds. Thus a continent *cannot* be easily transformed into an ocean, or vice versa. Wegener emphasized this point and thus refuted the land-bridge theory. Modern seismic and gravity studies of the ocean floor show that Wegener was right.

## Direct Linkage

The most convincing evidence of direct linkage between the continents is the distribution of ancient glaciers. Glaciation occurs at irregular intervals on the earth. In the present Quaternary period, which has lasted about two million years, the earth has passed through several glacial periods separated by interglacial periods. During the last such Ice Age, which ended only about 10 thousand years ago, most of Europe and North America were under thick layers of ice. During the period preceding the Quarternary period, however, the earth had been free from glaciation for more than 100 million years. Why glaciations occur only at certain times is still unknown and provides an interesting topic for debate, but one that space does not allow us to examine here.

What concerns us at this point is the type of evidence glaciers have left in the course of the earth's history. A thick continental glacier scrapes against the rocks as it moves, leaving unique traces called glacial striations; along the way it carves such topographic features as steep-walled glacial valleys. It also crushes and grinds up rocks, transports these fragments downstream, and deposits them at the front of the glacier as it melts. The resulting sedimentary deposits are so characteristic that their glacial origin can be recognized by the trained eye of the geologist even though millions of years may have elapsed since the glacier's melting. Examination of glacial distribu-

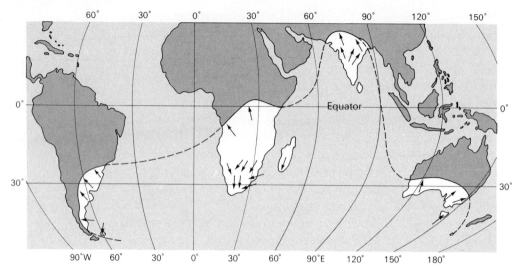

FIGURE 1-6
Map showing the distribution of the late Carboniferous glaciations of Gondwanaland with the continents in their present positions: arrows indicate directions of ice flow. [After A. Holmes, *Principles of Physical Geology*. Thomas Nelson and Sons, Ltd., Middlesex. The Ronald Press Company, New York, 2nd ed.; copyright © 1965.]

tion in the earth's ancient geological history reveals that glaciation was extensive in the Permo-Carboniferous period, approximately 300 million years ago. This glaciation affected all the continents in the Southern Hemisphere. If we look at a map of this glaciation (Figure 1-6), something about the distribution of the glaciers immediately strikes us: tropical regions such as India and Africa were under ice, but there is hardly any trace of glaciation in the rest of the Northern Hemisphere during this period, even on land masses near the present North Pole.

The theory of continental drift provides us with a clear-cut explanation. Figure 1-7 shows the original continent of Gondwanaland.★ Note that the glacial areas form a nearly circular icecap over the polar region of Gondwanaland. It was probably because of this impressive evidence that the continental drift theory attracted enthusiastic supporters from the Southern Hemisphere, such as A. L. Du Toit of South Africa and S. W. Carey of Tasmania, even after the theory was

---

★Gondwanaland was the giant continental mass of the Southern Hemisphere, consisting of the present southern continents. Its previous existence has been inferred from the distribution of such fossils as the Carboniferous flora *Glossopteris*. Taking the land-bridge theory as a basis, the 19th-century Austrian geologist E. Suess proposed the name Gondwanaland—the Gonds are an Indian aboriginal tribe.

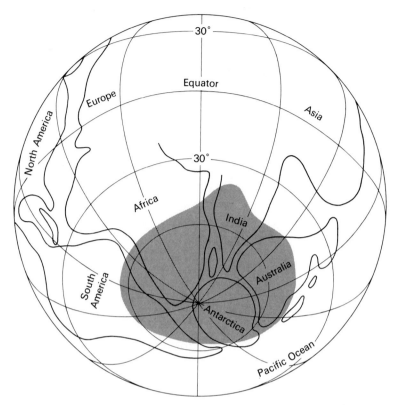

FIGURE 1-7
Map showing the distribution of the late Carboniferous glaciations of
Gondwanaland with the continents reassembled, as interpreted by A. Wegener.
[After A. Holmes, *Principles of Physical Geology.* Thomas Nelson and Sons,
Ltd., Middlesex. The Ronald Press Company, New York, 2nd ed.; copyright ©
1965.]

virtually abandoned by the majority of geoscientists of the Northern
Hemisphere. The geologists from the Southern Hemisphere had
themselves seen the ancient glacial traces, and knew they could be
explained only if it could be proved the continents had moved.

## The Contraction Theory
## and Continental Drift

As we saw in the introduction, one of the major geophysical issues
enumerated by Dr. Adams was that of *orogenesis*, or the origin of
great mountain ranges. How did those towering mountain ranges—

the Alps, the Himalayas, the Rockies, and the Andes—originate? Some of the thick strata that form these mountains are sediments deposited on the sea floor long ago, indicating that such mountains were somehow elevated from that floor. It is awesome to think that where a mountain now towers there once existed an ocean basin in which a layer of sediments, 10,000 meters thick, was deposited. Yet many such basins, called *geosynclines*, have formed on the sea floor, only to be lifted up later to form mountains. What could account for such an upheaval?

The previous leading theory on orogenesis was predicated on the notion of a contracting earth. It in turn was based on another theory, the "hot-origin" hypothesis of the earth, which assumed that the planet was once a ball of "fire" or incandescent gas, which subsequently condensed and gradually cooled down. Most geological surveys of mountain ranges reveal a tilting of the strata. In many areas the strata are bent into a wavelike pattern called *folding*, which consists of alternating arches (*anticlines*) and troughs (*synclines*); in some formations, the folding is so extreme that the strata are upside down as in Figure 1-8. The contraction theory seemed to explain this phenomenon of folding. According to the theory, as the surface of the hot earth began to cool, solidify, and contract, its volume decreased. The interior, however, was still hot. Because of the tension produced in the rapidly shrinking outer layer, cracks began to form on the surface, just like the cracks that form in drying mud. Geosynclines might have occurred within such giant cracks, where water could gather and where deposition could take place. The interior too would eventually cool, so that its volume would decrease. This contraction would begin to exert compression on the already cooled surface. Like a suit of clothes that is too large, the crust was now too big for the shrinking interior, and wrinkles formed. This was the explanation for folded mountain ranges.

The theory sounded plausible but it had yet to be proved quantitatively. In an attempt to do so, geologists first determined the degree to which strata in mountain ranges had been compressed. From these results it was calculated that the entire earth had to have cooled by thousands of degrees to produce enough contraction to form a single mountain range several thousand meters high. Such extreme cooling seemed unlikely. To complicate matters further, each mountain range was formed at a different time, some of them quite recently. It seemed impossible that the earth could have cooled by thousands of degrees for each of them. This problem had already been pointed

FIGURE 1-8
Example of sharply folded strata of Tertiary bedded sandstone, Kii Peninsula,
Japan. [Photo by F. Kumon, Kyoto University.]

out in Wegener's day, and since then the contraction theory has
lost ground.

Today, as a consequence of various cosmological studies, even the
basic assumption of an earth that has cooled from an incandescent
hot state is in serious doubt. So the once prevalent contraction theory
no longer seems like a plausible explanation of the origin of mountains. Wegener declared that no contraction was necessary to produce
the folding of strata and the formation of mountains. He contended
that the leading edge of an advancing continent would encounter
resistance and, as a result, compress and fold. As North and South
America drifted westward, leaving the Atlantic Ocean in their wake, a
chain of mountain ranges formed along their leading edge; the Sierra
Nevada and adjacent mountain ranges in North America and the
Andes in South America. Wegener further suggested that when
Gondwanaland split, India drifted northward and eventually collided
with the Asian continent. The overriding of India by Asia in the zone
of collision caused the Himalayas to form.

Meteorologist Wegener's continental drift theory was a breakthrough in the complex field of orogenesis, which for years had been

the awkward and misunderstood stepchild of the professional geologist. The challenge posed by Wegener's bold theory was welcomed by some geologists, but the majority were skeptical of such simple logic.

As already mentioned, many of the strata forming today's mountain ranges originally accumulated under the sea in thicknesses often exceeding 10 thousand meters. All such strata, the scholars of that day agreed, had been deposited in shallow water. But if the sea was shallow, how could such thick deposits have accumulated? The only explanation seemed to be that the sea floor had sunk as more and more layers of deposits accumulated, so that the ocean depth remained constant. Thus, the deposits sank deeper and deeper into the earth. Today, the same strata rise high above sea level. At some point the process of depression of the sedimentary basin, or geosyncline, must have somehow reversed so that the strata were thrust upward into mountains. Why did the basins form in exactly those places that subsequently became mountains? We see that for this problem too Wegener's theory provided the key.

## What Moved the Continent?
## A Challenge to Geophysics

The theory of continental drift presented a challenge to classical geology because it provided a simple and logical explanation for so many geologic processes. However, it posed an even greater challenge to geophysics.

The question it raised was basic. What kind of force could cause the continents to move distances of several thousand kilometers? What was the driving mechanism of continental drift? That is, an explanation of *effect* was meaningless if the *cause* could not be identified: so even if Wegener's theory of continental drift did provide lucid explanations for many geological effects, such as ancient glaciation and mountain building, it could scarcely be regarded as scientific unless it could also explain what had originally caused the continental movements. Although he knew how crucial this initial cause was to his theory, Wegener never succeeded in explaining it. It was not easy, even for a man such as Wegener, to move the "immovable" earth.

As Figure 1-1 shows, Wegener proposed that Gondwanaland was first located around Antarctica. Later it began to split and left the South Polar region. This made Wegener suspect that continents in general move away from the poles and drift toward the equator. He named the force behind this phenomenon the *pole-fleeing force*, and explained its origin as follows: Because the earth rotates on its axis, there is a centrifugal force of rotation. This force deflects the pull of gravity slightly so that it is directed *not* toward the center of the earth but toward the equator, though only very slightly. Consequently, Wegener reasoned, continents floating on the earth gradually move toward the equator. This hypothesis was barraged by objections founded on the actual magnitude of the force; as computations made by various scientists showed, the pole-fleeing force is extremely small. In fact, it is several millions of times smaller than the force of gravity. But Wegener insisted that, however small it may be, a force acting continuously for a long time can, in the end, move a continent. Many others continued to believe, however, that a far greater force would be necessary to displace the continents floating upon the solid mantle. Moreover, the force would have to be great enough to fold and hoist once flat layers of sediments into mountains several thousand meters high—an impossible task for the pole-fleeing force.

The westward drift of the two American continents Wegener attributed to the tidal attraction of the sun and the moon. This theory, too, was refutable. The famous British geophysicist, Sir Harold Jeffreys, led the opposition. In each edition of his well known and sophisticated book *The Earth*, Jeffreys (1970) criticizes Wegener's theory of continental drift as theoretically impossible.

## The Death of the Theory

Until the late 1920s, Wegener's theory of continental drift remained the subject of heated controversy. Then interest declined almost completely—first, because the theory contradicted the commonly accepted belief of the time that the earth was solid and hard, and, second, because Wegener had failed to provide a satisfactory explanation for the force that had set the continents in motion. Scientists could not accept Wegener's supposition that continental drift had occurred in the earth's recent history—say, the last 200 million years, which constitutes only a very small percent of the earth's total age of 4500 million years. They reasoned that if continental drift were possible at all, it could have occurred only during the earlier part of the earth's history when it was still hot and soft. Their reasoning was of course based on the "hot origin" hypothesis, the dominant theory of the time. Thus the theory of continental drift, born in 1912, was virtually dead by the 1930s. Only a few diehard supporters remained—a handful of geologists of the Southern Hemisphere who were still faced with the fact of the Permo-Carboniferous glacial distribution among the continents in that part of the earth.

## The Structure of the Earth

Abandoned by most scientists for twenty years or so, the theory of continental drift experienced a dramatic comeback in the late 1950s. Strengthened by new evidence, it is currently forcing us to change our view of the earth. This revival will be more clearly understood if we summarize a few additional basic facts about the earth. First, the structure of the earth's interior is studied by seismological methods. These methods are somewhat analogous to tapping on a watermelon to see if it is ripe. Like the sound we get when we tap the watermelon, seismic waves reveal the internal state of the earth. They tell us the earth is layered like an onion, consisting of the exterior crust, a solid mantle that extends to a depth of 2900 kilometers below the surface, an outer core believed to be liquid, and finally a solid inner core about 1100 kilometers in radius at the center of the earth (see Figure 1-9). The crust consisting of the lighter-weight granite and other rock types, has a low density. Each successive layer has a higher density than the one above it. The mantle, beginning at the Mohorovičić discontinuity, consists of heavier rocks such as peridotite. Until recently it was thought that the mantle was entirely solid, but we now suspect that at certain depths the rocks are so close to their melting point that they are able to flow plastically. The outer layer of the core is generally thought to consist mainly of liquid iron mixed with such elements as nickel, carbon, silicon, or sulfur. (This liquid state of the outer core has a direct bearing on the earth's magnetism, as we shall see in the following discussion.) The inner core is composed of the same elements in their solid form.

## The Earth's Magnetism

As everyone knows, a compass needle invariably points to the north or nearly so. As early as the 14th century, sailors were using this phenomenon for navigational purposes. William Gilbert (1600), physician to Queen Elizabeth I, explained the phenomenon by proposing that the earth itself was a huge spherical magnet with its poles

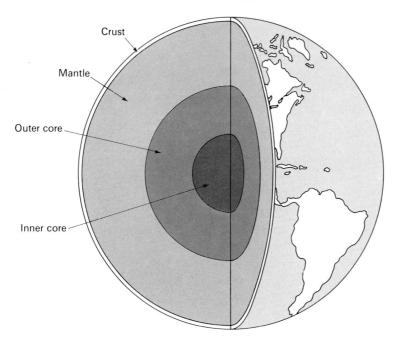

FIGURE 1-9
Cross section of the earth. Note that the thickness of the crust is exaggerated.

situated almost at the geographical poles as shown in Figure 1-10. If
so, and because unlike magnetic poles attract each other and like
ones repel, magnetic compass needles would naturally tend to point
one end to the north and the other to the south. This was Gilbert's
insight. The N and S poles of a magnet should thus be called, more
properly, the north-seeking and south-seeking poles. (It is interesting
to note that the earth's magnetic pole at the geographic north pole is

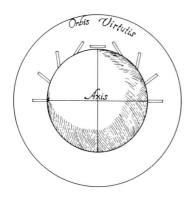

FIGURE 1-10
Spherical magnet earth of W.
Gilbert. [After W. Gilbert, *De
Magnete, Maneticisque Corporibus,
et de Magno Magnete Tellure
Physiologica Nova.* Short, 1600;
Dover, New York.]

in fact a magnetic south-seeking pole; it has to be in order to attract the north-seeking pole of a compass.)

Gilbert's explanation was right but—as always happens in scientific research—it produced another question: why is the earth a magnet? This is a tough one. Recall that the origin of the earth's magnetic field was also listed by Dr. Adams in 1947 as one of the most difficult unsolved problems in geophysics. The simplest of all theories of the origin of geomagnetism was the assumption that the center of the earth was a huge permanent magnet. It was well-known that, among the common metals, only iron and nickel could be permanent magnets. (Such materials are called *ferromagnetic*.) Since the earth's core consists mainly of iron and nickel, the explanation of the earth's magnetism seemed obvious. This assumption, however, tunred out to be wrong for a simple reason. All ferromagnetic substances lose their ferromagnetism when heated beyond a certain temperature. That is, a magnet does not remain a magnet once it has reached a certain temperature, which is called the *Curie point* (770°C for iron and 358°C for nickel). It was then evident that the iron and nickel in the core could not form a permanent magnet since the temperature in the core was certainly higher than the Curie points of either metal. Since the outer part of the earth's core is liquid as revealed by seismic waves, its temperature is obviously higher than the melting point of iron and nickel, and laboratory experiments demonstrate that the Curie points of iron and nickel are much lower than their melting temperatures. In fact, it is only about the outer 50 kilometers of the earth that is cool enough to permit any material to be ferromagnetic.

Another hypothesis was that any rotating body was inevitably magnetized as a consequence of its rotation. The late English Nobel laureate physicist P. M. S. Blackett, who proposed the hypothesis, pointed out that the magnetism of such celestial bodies as the sun, certain stars, and the earth, could all be explained as being due to their rotations. He emphasized that such an explanation was not founded on the established laws of physics but required the positing of an entirely new concept. Blackett set out to prove his theory by developing an amazingly precise magnetometer in the late 1940s. However, his efforts failed, disproving his own hypothesis. Fortunately, in reporting his failure, he gave a remarkably full description of his sophisticated and precise measurements in a well known article entitled, "Negative Experiment" (1952). Indeed, in a deeper sense, the experiment was not a failure because of the extraordinarily sensitive magnetometer he developed to make the experiment. This magnetometer later proved to be a most useful tool when Blackett began to study the magnetism of rocks, and these experiments made a vital contribution to the revival of the theory of continental drift. Models of Blackett's magnetometer are now found in many geophysical laboratories all over the world.

Of the many theories that have been proposed, only one explanation of the origin of the earth's magnetic field has survived—that in which the earth is viewed as an electromagnet rather than as a permanent magnet. A magnetic field can be generated either by a permanent magnet made of ferromagnetic minerals or by an electric current. In the 1950s such scientists as W. M. Elsasser of the United States and Sir Edward Bullard of England concluded that, since the earth was too hot to be a permanent magnet, it must be some sort of electromagnet, and they began to explore vigorously the possibility that a geomagnetic field was produced by electric currents in the earth.

In order to provide enough flow of electricity to create the geomagnetic field, the electric conductivity of the earth's interior would have to be as high as that of metal. The iron core is the only part of the earth that could possibly have such a high electrical conductivity. In addition, an electromotive force or voltage must be constantly present to keep the electric currents flowing and maintain the geomagnetic field for a geologically long time. In other words, the earth's core has to be more than a good electrical conductor through which electric currents pass: it must also act as a dynamo or generator. This concept of the origin of geomagnetism is called the *dynamo theory*.

However it is almost inconceivable that a mechanism like the generators we are accustomed to—complicated pieces of machinery with insulated wires—would exist inside the earth's core. Yet in the 1960s, young geophysicists from the United States (G. Backus) and England (A. Herzenberg) proved that it was possible, at least theoretically, that a body like the earth's core could act as a dynamo. In 1963 F. J. Lowes and I. Wilkinson of England succeeded in constructing a generator somewhat similar to the one described in the theory. H. Takeuchi, Y. Shimazu and T. Rikitake of Japan also contributed to the development of this model. However, the theory has not yet been completely established. As electronic computers have become more and more powerful, the theoretical calculations have also become more and more sophisticated, and additional complications in the theory have been disclosed. Perhaps it is accurate to say that the origin of the earth's magnetic field remains a great mystery. But theoreticians are willing to concede at least the possibility that the earth could have a dynamo-generated magnetic field.

The dynamo theory as it stands today assumes an extremely complex chain of processes taking place in the earth's core. We will not describe all of these processes here, but it is important to recognize that the following conditions are necessary if the earth is to work as a generator:

(1) the core of the earth must consist of a substance that conducts electric current as easily as metal does;

(2) the substance must be in a liquid form;

(3) this conducting liquid must be stirred up in some way, the stirring process providing the energy needed to sustain the field.

These conditions make it almost imperative that the core of the earth consist of liquid metal that is probably iron—the most common and abundant metal in the universe.*

## Paleomagnetism: The History of the Earth's Magnetic Field

At first glance, continental-drift theory and geomagnetism seem to have little in common. Yet it was shown in the late 1950s that the two are actually quite closely related. The first step in this recognition occurred when geophysicists questioned whether the compass needle always pointed to the north. If the geomagnetic field is produced by a "permanent" magnet, the history of the earth's magnetism must have been a boring record of consistency; but if we consider geomagnetism as an electromagnetism produced by a "dynamo," its history can vary a great deal. The dynamo concept was supported by the following observation. In present-day Tokyo, the compass needle deviates 6° to the west of exact north. The angle of deviation is called

---

*The famous American geophysicist F. Birch wrote in his classic paper (1952) the following:

    Unwary readers should take warning that ordinary language undergoes modification to a high-pressure form when applied to the interior of the Earth; a few examples of equivalents follow:

| High-pressure form | Ordinary meaning |
|---|---|
| certain | dubious |
| undoubtedly | perhaps |
| positive proof | vague suggestion |
| unanswerable argument | trivial objection |
| pure iron | uncertain mixture of all the elements |

the *declination*. An interesting fact is that the declination changes in the course of time; 150 years ago the compass needle in Tokyo was recorded to have pointed 3° to the east. This phenomenon is widely known throughout the world and is called the *secular variation* of geomagnetism. The actual measurement and recording of geomagnetism began only 300 years ago. Naturally this is too short a period to provide information about the changes that have taken place throughout geological time. It is important, however, to note that even within the span of human history, geomagnetism has changed considerably. How much greater might those changes have been throughout the long span of geological time? Answers to this interesting question may reveal the nature of geomagnetism. But how are we to study geomagnetism as it was a million years ago? The magnetic field is just a "field" that, if it changes with time, leaves no indication of its former condition—it is therefore extremely difficult to trace its history. Nevertheless an interesting possibility was discovered: the permanent magnetization of natural rocks sometimes provides us with a "fossil" that contains a trace of the magnetic field as it once was. (Among the various kinds of natural rocks, let us consider volcanic rocks—which are cooled and solidified magma. An examination of volcanic rocks such as basalt reveals surprisingly strong magnetism. Of course its strength is but one thousandth of that of the usual magnet that can attract iron and suspend nails. Yet, with a sensitive device, it is a reasonably easy task to determine the direction of magnetization of volcanic rocks. But why are they magnetized in the first place? The answer is as follows: when a volcanic rock comes into existence, that is, when it is erupted from a volcano, it is incandescent lava and its temperature is much higher than the Curie point. As the lava cools through the Curie point its magnetic moment is set in the direction of the geomagnetic field at that time and remains in this condition permanently. This "fossilized" magnetization is characteristic of all volcanic rocks. It was investigated by, among others, J. G. Königsberger of Germany, T. Nagata of Japan, and E. Thellier of France, in the 1940s. The famous French Nobel laureate L. Néel provided an ingenious theoretical explanation to this pehnomenon, called *thermoremanent magnetization*. Once the mechanism of such magnetization had been established, it became possible, at least in theory, to trace the history of the earth's magnetic field by measuring the magnetization direction of rocks from various geological periods.

This field of study is called *paleomagnetism*. Paleomagnetism rose in popularity in the 1950s and disclosed many new facts, the most remarkable among them being those related to the revival of the continental drift theory! Before treating this important aspect of paleomagnetism, however, let us digress to examine another issue that is equally important to the theme of this book.

### Reversal of the Earth's Magnetism or Self-Reversal of Rock Magnetism?

The study of paleomagnetism has traditionally been very active in France and Japan. B. Brunhes and M. Matuyama were the foremost pioneers in this field. Brunhes discovered, as early as 1906, that some rocks are magnetized in the opposite direction to the present geomagnetic field and proposed the possibility that the earth's magnetic field had been reversed when these rocks were formed. Matuyama in the 1920s found that about half the volcanic rocks from Japan and Korea that he measured were magnetized in the same direction as the earth's magnetic field at present. But the other half were magnetized in the opposite direction. On the basis of this study Matuyama concluded that the earth's magnetic field had reversed near the beginning of the Ice Age, in the early Pleistocene. This was a bold assertion at the time. Late in the 1950s, however, the same evidence turned up repeatedly in Iceland, France, England, the United States, the USSR, and elsewhere. A. Cox, R. Doell, and B. Dalrymple of the United States and I. McDougall, D. Tarling, and F. Chamelaun of Australia investigated this issue thoroughly and established the reversal history of the geomagnetic field for the last several million years.

I must confess that I too have a deep personal interest in this problem. In 1951, when I was an undergraduate student at the University of Tokyo, I was performing a series of experiments under the guidance of T. Nagata. I was examining the way in which the ferromagnetic minerals that are contained in various volcanics acquire thermoremanent magnetism when cooled through the Curie point in a magnetic field. The procedure was to heat the samples, which were contained in silica glass tubes, to above the Curie point and then cool them in a magnetic field. In the course of these experiments, I noticed that one of the samples, which consisted of ferromagnetic grains extracted from a pumice of the Japanese volcano Haruna, had been magnetized in the opposite direction to that of the applied magnetic field in my laboratory. Such an observation could have been the result of my having mismarked the orientation of the sample; certainly the acquisition of magnetization in a direction opposite to an applied field appeared impossible or totally absurd as long as the fundamental laws of physics held true. But the observation was unmistakably real. Being a lazy student, I did not trouble to repeat the heating and cooling experiments for each sample, but instead had put several samples together in a furnace. Therefore, when I found that only one of them had been magnetized in a direction opposite to all the others, I knew there was no chance of a mistake. Both my professor and I were completely perplexed by this odd phenomenon. But before long we realized that it could be an important discovery. We avidly conducted various experiments and devised a "theory" to explain the physical cause of this phenomenon of *reverse thermoremanent magnetism*, as we named it. About that time, T. Rikitake drew our attention to a paper by L. Néel (1951), in which such a phenomenon was theoretically predicted. The paper had been published in France at about the same time we were discovering the phenomenon in Tokyo. We were impressed by his insight. Later we learned that Néel's work had been inspired by an American geologist John Graham, who had written a letter to Néel asking if such a phenomenon might be theoretically possible. In fact, what prompted Graham's question was the frequent natural occurrence of rocks that are magnetized in a direction opposite to the present geomagnetic field. Instead of assuming the reversal of the geomagnetic field, he wondered if some rocks might have an intrinsic property of reverse magnetization! John Graham died in 1971, but will be long remembered for his imaginative and ingenious ideas on many aspects of earth science.

Upon the discovery of the self-reversal of remanent magnetization in rocks, some scientists, myself included, suggested that we need not assume the reversal of the geomagnetic field in the geological past. In fact, for several years papers on this subject flowed in continuously from numerous parts of the world. These were all studies of rocks that were naturally magnetized in the opposite direction from the earth's magnetism, and the objective was to see whether they, like the rock from Mt. Haruna, could also self-reverse the direction of magnetization. The results of the papers proved, contrary to our expectation, that such rocks were quite rare. Although self-reversal was found to be an uncommon occurrence requiring a very special kind of ferromagnetic mineral, this particular kind of thermoremanent magnetization was such a fascinating phenomenon that I devoted myself to a quest for its mechanism for a good six years. The ultimate cause of this phenomenon was found to be different from the original models proposed by either Néel or us. Rather, it appears to be related to highly intricate quantum-mechanical interactions taking place within minerals contained in the sample. The problem is today still the subject of investigation by young scientists around the world.

The evidence, then, is so overwhelming that we must, at least for the present, concede that the earth's magnetism *did* reverse frequently during the geologic past. The significance of the reversal of geomagnetism to the new view of the earth will be made clear in Chapter 2.

### Poles Move and So Do Continents

If we take a sample of lava from Japan's Mt. Fuji and measure the direction of its remanent magnetism, it is possible to deduce the position of the earth's magnetic pole when that lava poured forth. The earth's magnetic field can be approximately represented by a regular dipolar pattern (Figure 1-11) that closely resembles the field produced by a bar magnet placed at the earth's center. This pattern enables us to determine the position of the South and North Poles by examining the direction of the magnetic lines of force at any given location. For a variety of reasons, however, the earth's magnetic field does not form a perfect dipolar pattern. The pattern is actually far more complex. Thus no accurate position of the pole can be calculated if the calculation is based on the assumption of a perfect dipole field. Nevertheless such computations are approximately correct—especially when a sufficiently large number of measurements are made and their average is taken—as has been demonstrated by current measurements of many rocks of recent age all over the world. If we assume the field was also a dipole in the past, the positions of the geomagnetic pole in earlier ages may be estimated from measurements of the direction of the natural remanent magnetism of older rocks. This assumption—that the earth's magnetic field was always dipolar—is an important one, but it is still only an *assumption*.

The study of paleomagnetism developed mainly in Japan and France during the late 1940s and early 1950s. Then in the mid-1950s it was taken up by British scientists who applied it with skill and enthusiasm to the examination of rocks of many ages from all over the world in order to survey extensively the history of the earth's magnetic field. Led by P. M. S. Blackett and S. K. Runcorn, they exerted an enormous effort in this work. It is a well known fact that the highly sensitive magnetometer developed by Blackett for his "negative experiment" (described earlier in this chapter) was of great

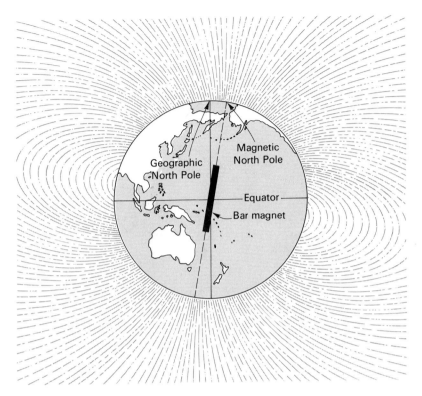

FIGURE 1-11
The earth's magnetic field is much like the field that would be produced if a giant
bar magnet were placed at the earth's center and slightly inclined (11°) from the
axis of rotation. [After F. Press and R. Siever, *Earth*. W. H. Freeman and Company,
San Francisco. Copyright © 1974.]

assistance in these projects. The British enthusiasm for this seemingly unexciting task at that particular time puzzled those of us working in Japan. The English are said to have an unconditional love of
nature and the earth—perhaps this was the reason, but the discerning
scientific leadership of Runcorn and Blackett seems to have had a
significant influence as well. Although the British scientists might
have appeared to be engrossed simply in examining the magnetism of
rocks, their efforts must have been motivated by a great deal of
foresight on someone's part. Whatever the impetus, they scattered
themselves throughout the world, collecting and examining rocks.
By 1957 they had achieved brilliant results.

Runcorn and his group attempted to represent the earth's magnetic
field in the past by the position of the ancient magnetic pole. This
was the best possible way to analyze uniformly the results obtained

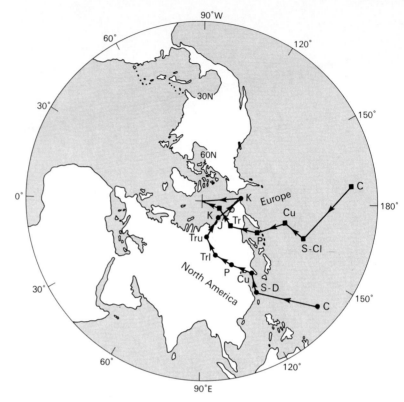

FIGURE 1-12

Comparison of the apparent polar-wandering paths for North America (circles) and Europe (squares). The circles and squares themselves represent the essentially stable regions of each continent for the different geologic periods. The following letter symbols are used to designate the various periods: K—Cretaceous; Tr—Triassic; Tru—Upper Triassic; Trl—Lower Triassic; P—Permian; Cu—Upper Carboniferous; S-D—Silurian-Devonian; S-Cl—Silurian Lower Carboniferous; C—Cambrian. [After N. W. McElhinny, *Paleomagnetism and Plate Tectonics*. Cambridge University Press, 1973.]

from rocks collected from locations that were such large distances apart. First the rocks from England and the European continent were examined to determine the position of the magnetic pole in each geologic period from the Precambrian era to the present. The result is plotted on the map in Figure 1-12. If the magnetic pole has not changed its position throughout the earth's history, the plot will point to a single spot. However the result conclusively shows a systematic movement of the pole. About 250 million years ago during the Permian period, the magnetic pole was located north of where the Japanese islands are at present—quite a distance away from the present North Pole. Five hundred million years ago, during the Cambrian period, it was much farther away in the Pacific. This phenomenon is called *polar wandering*.

Interestingly enough the path of polar wandering, as determined paleomagnetically, roughly corresponded to the path of yet another pole traced by an entirely different method. This other pole was the paleoclimatological pole, which was found by the location of fossils of ancient plants and animals. This observation implied that the various locations of the ancient paleomagnetic poles were indicative of changes in the orientation of the earth's rotation axis. The underlying assumption was that, in the past, the polar regions had been cold and the equatorial regions warm, and that the fossils of life forms indicate the former latitude of localities at which they are sampled today. Based on this kind of analysis, such scientists as Wegener, Köppen and Kreichgauer had already talked about polar wandering as early as the 1910s. The coincidence of the paths of the magnetic pole and that of the paleoclimatological pole, though far from exact, must have encouraged the British scientists.

Runcorn and his group conducted their intensive search for the path of the magnetic pole, using rocks not only from Europe but from North America as well. The result is also shown in Figure 1-12. To be noted are the loci of the paleomagnetic poles as estimated from the rocks of England and Europe and from those of the North American continent. Anyone can see that the paths for each are similar, and form such a coherent pattern that it is hard to dismiss polar wandering as false or accidental. Furthermore, careful scrutiny will reveal that these two lines, although they are quite similar, are *not* identical. The discrepancy appears to be systematic. S. Runcorn and E. Irving, examined this discrepancy closely, and came up with an idea that was to revive the theory of continental drift.

Their idea was a simple one. If paleomagnetists had been alive during the Permian period, 250 million years ago, they would have found that paleomagnetic poles for rocks forming at that time at different sites all over the world would coincide, just as they do today—in particular, those from Europe and North America. Now let us see what happens if North America moves away from Europe. A paleomagnetic pole behaves as if it were attached to a continent by a rigid rod because it is determined from paleomagnetic measurements, which tell us that the ancient pole was at a certain distance along a specified great circle from a sampling site. If a continent moves, the pole moves with it. So if North America has moved away from Europe since the Permian period, its pole has moved with it,

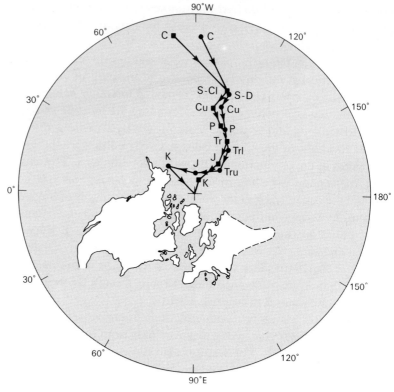

FIGURE 1-13
The two polar-wandering paths in accord with the fit of the North Atlantic
proposed by Bullard and others. As in Figure 1-12, the circles (North America) and
squares (Europe) represent the stable regions of each continent for each period.
K—Cretaceous; Tr—Triassic; Tru—Upper Triassic; Trl—Lower Triassic;
P—Permian; Cu—Upper Carboniferous; S-D—Silurian-Devonian;
S-Cl—Silurian Lower Carboniferous; C—Cambrian. [After N. W. McElhinny,
*Paleomagnetism and Plate Tectonics.* Cambridge University Press, 1973.]

and we can no longer expect the two to coincide. Runcorn and Irving
found that the poles for North America and Europe were distinctly
different, demonstrating that the two continents had moved apart.
They also found that if they pursued Wegener's ideas and closed up
the Atlantic to restore the continents to their former position, the
magnetic poles coincided, as illustrated in Figure 1-13. Their research revived the theory of continental drift and provided completely
independent evidence in support of it.

During my stay at Cambridge University from 1958 through 1959,
nearly every time I was introduced to a geophysicist, I was greeted
with such questions as "Do you believe in continental drift?" or "Do

you believe in the reversal of the earth's magnetic field?" Being a little unfamiliar with the inclination of British scientists to favor such ideas as continental drift, my answer used to be a half-hearted "Well, yes, but with reservations." I was aware of what was being discovered at the time, and I knew the evidence was quite solid, but my enthusiasm could not quite match that of the English.

As I've already mentioned, why the British should revive the continental drift theory at that particular time through their ardent studies of rock magnetism seemed a mystery to others of us. Scientists who had been active when the theory of continental drift was still popular were already too old to be acquainted with the new field of paleomagnetism, and yet in most countries the younger scientists who explored this new field of paleomagnetism were not really familiar with the theory of continental drift. England, however, was probably an exception, owing largely, I believe, to the excellent textbook *Principles of Physical Geology* (1965) by the late Edinburgh University professor, Arthur Holmes. In this book, the then unpopular theory of continental drift was still vividly discussed along with Holmes' famous theory of convection in the mantle.

## The Theory of Convection in the Mantle

As we have seen, the theory of continental drift was abandoned in the 1930s because no satisfactory explanation of what causes continental movement was produced. Of the many hypotheses suggested by Wegener but not developed by him rigorously, only one has survived: that the mantle undergoes thermal convection similar to that seen in a kettle of soup on a stove: as the soup at the bottom is heated it expands, becomes less dense, and rises to the top. At first the proposal that a similar process takes place in the earth might seem absurd because the mantle is solid. However a number of materials, like cool tar and "silly putty" will break like a solid if they're bent quickly but will flow slowly like a liquid if gentle forces are applied to them over a long period of time. During very long periods of time even ice is able to flow plastically, and so can the mantle. Arthur Holmes maintained it was this flow due to convection that provided the driving mechanism for continental drift. He likened the flow to, in his words, "an endless travelling belt"—or what we today refer to as a "conveyor belt"—and asserted that even a continent could be carried along by it.

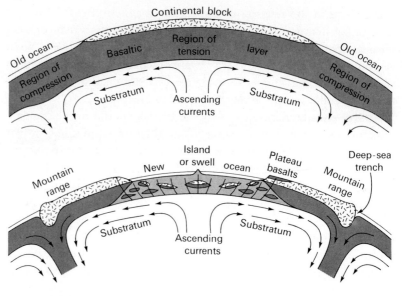

FIGURE 1-14

Model demonstrating mantle convection as a possible mechanism of continental drift. It shows a continent being pulled apart by rising mantle currents, with new ocean developing from the growing rift. In the vicinity of a descending current, a mountain range and bordering deep-sea trench develop. [After A. Holmes, *Principles of Physical Geology*. Thomas Nelson and Sons, Ltd., Middlesex. The Ronald Press Company, New York, 2nd ed.; copyright © 1965.]

All of the other theoretical mechanisms for continental drift had been founded on the fixed idea that the continent itself pushes its way through the solid mantle. Once theoretical studies conducted by geophysicists had shown this propulsion was impossible, geophysicists lost interest in the continental drift theory.

Holmes theorized that if the flow within the mantle welled up in the middle of a continental mass and parted to each side, the continent would split and the two halves would drift apart. The Atlantic Ocean has formed in such an expanding rift. The mechanism of this theory is represented in Figure 1-14.

Holmes' theory, which he first proposed in 1929, survived without receiving much active opposition, probably because his ideas were too far ahead of the times. If we examine the model in Figure 1-14 closely, we cannot help but recognize its striking affinity to the new view of the earth—the sea-floor spreading hypothesis, to be described in later chapters. We will refer to the problem of the convection in the mantle elsewhere in this book.

British scientists continued their remarkable exploration of this fundamentally revolutionary concept of earth's science until the end of the 1950s. Then, in the 1960s, the scientists of the new world arrived on the scene.

## A Modern Jigsaw Puzzle

Wegener first conceived the idea of continental drift upon trying to fit together the two Atlantic coastlines. This method was later extended to establish conformity between the fossils of ancient plants and animals and between the geologic strata of each continent. Some scientists, like Wegener, found that continental conformity was very close, whereas others reported large gaps and areas of overlap. All the attempts to assess the fit one way or the other were criticized as too subjective. In recent years, more objective methods have been developed.

Sir Edward Bullard and his colleagues (1965) settled the argument by programming an electronic computer to try all possible rearrangements and find which fit the best. They discovered that the contour line at a depth of about 1000 meters, rather than the present coastline, fitted best (Figure 1-15). They made the quite reasonable suggestion that this line be considered as the contour of the original continent. The fit determined by the computer is amazingly good. Although overlaps and gaps do exist, they are extremely small. It would thus seem that the electronic computer has proved that the continents fit together almost as perfectly as the pieces of a puzzle. Assuming that this conformity was not purely coincidental (as is the boot shape of the Italian peninsula for example), Bullard concluded that it strongly suggested that these continents originally formed one continental mass.

Another method was developed by P. Hurley (1968) of the United States and his colleagues from Brazil. A number of geological features on both sides of the Atlantic match, but it was always suspected that an element of subjectivity had influenced the matching process. These workers made it possible to see how well the continents fit together objectively by determining the absolute ages of rocks with the radiometric method, which is almost entirely objective. In this method, as explained near the beginning of this chapter, the rate of spontaneous disintegration of radioactive elements with long half lives is the basis for determining the age of rocks with as much certainty as can be provided by modern physics. With this technique,

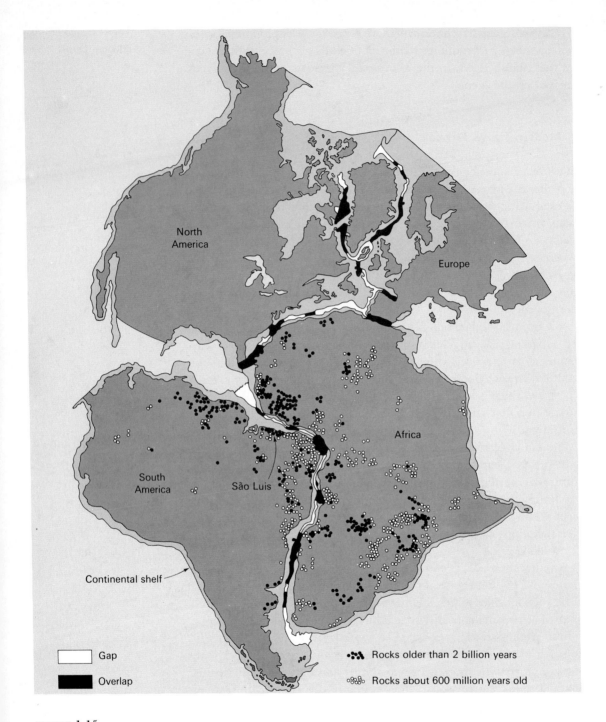

| | |
|---|---|
| Gap | Rocks older than 2 billion years |
| Overlap | Rocks about 600 million years old |

FIGURE 1-15
Map showing the conformity of the continents bordering on the Atlantic. The black areas along the coastlines represent the continental overlaps, and the white areas the gaps. Also matched are the ages of the rocks in South America and Africa. The dark circles denote rocks older than two billion years; the light circles denote the younger group approximately 600 million years old. [After P. M. Hurley, "The Confirmation of Continental Drift." Copyright © 1968 by Scientific American, Inc. All rights reserved.]

Hurley and others analyzed the ages of enormous numbers of ancient rocks from the eastern part of South America and from the western part of Africa, as shown in Figure 1-15. The dated rocks of the two continents fall neatly into two groups: those more than two billion years old and those approximately 600 million years old. Both ages are older than the proposed split of Gondwana. The regions of the same age were found to match across the Atlantic as they should. A typical example, pointed out in the figure, is the region considered to be originally a part of the ancient African continental block, now isolated on the coast near São Luis, Brazil. In view of these findings, it became increasingly difficult to dismiss the continental-drift theory as the wild idea of a meteorologist who dabbled in geology.

## Chapter 2

# The Exploration of the Ocean Floor

### A New Frontier: The Science of the Ocean Floor

While paleomagnetism was bringing about the dramatic revival of the theory of continental drift, rapid progress was being made in an entirely different field—the science of the ocean floor. The fact that the ocean covers two thirds of the earth's surface makes research on the ocean floor crucial to the understanding of the earth as a whole. Even before World War II, scientists had already begun to appreciate its significance. But, as the study of oceanography progressed, it became clear that it was not the great extent of the oceans that makes marine geoscience important. It was the distinct *nature* of the oceans that was significant.

Vening Meinesz of Holland was one who began to suspect this as early as the 1930s. He had developed a technique for measuring gravity at sea that was amazingly advanced for that time. The measurement of gravity requires a precision higher than one part in a million. It requires precise leveling of the gravity meter and is extremely difficult even on land. It was almost an impossible feat to take such a measurement in the ocean, where no fixed or stable station is available. But Meinesz set up his sophisticated gravity meter on a submarine that remained reasonably stable while submerged deep beneath the waves, and he proceeded to measure the gravity at many spots in the ocean. His survey was particularly thorough in the waters around the Indonesian archipelago, then

Dutch territory. This method of gravity measurement was also employed by M. Matuyama and others of Japan, using a submarine of the Imperial Japanese Navy. These surveys revealed the remarkable anomalies in the distribution of gravity around the deep-sea trenches along Indonesian and the Japanese island arcs. These anomalous zones of gravity were entirely new discoveries. They simply do *not* exist on land. Thus earth science had a hint that something utterly new was to be found in the ocean. From these results, Meinesz came up with a hypothesis on the origin of trenches and island arcs, based on the theory of convection currents within the mantle.

It was obvious to those who had "insight" that research on the ocean floor was indispensable to the solution of basic geological problems such as the origin of continents, the origin of oceans, and the structure of the mantle. Both the technology and the financial support for such research, however, were still very scarce at that time. Despite much endeavor by a few enlightened scientists in the 1930s and 1940s, the real development of ocean-floor science had to wait for the end of World War II.

## Prerequisites for Studies of the Ocean Floor

The survey of the ocean floor, which is thousands of meters deep, cannot be made with the ordinary techniques used in land geology and geophysics. One needs special tools. First a research vessel is necessary. It must be provided with various kinds of special equipment in addition to that required for long-distance navigation. For example, special devices are needed for the exact positioning of the vessel and for measuring the water depth; wires and winches long enough and strong enough to lower the various instruments to a depth of many thousands of meters are also needed. Besides, research of this kind is a time-consuming process and requires a vessel that is reserved uniquely for this particular type of investigation. For these reasons, oceanic research immediately after the war was undertaken mostly by the victorious nations, such as the United States, England, and the USSR.

Ironically, once the techniques and methods were developed and research vessels secured, research on the oceanic areas progressed much faster than that on land. Whereas survey equipment must be hauled across valleys and over mountains, sometimes from nation to nation, the vessels can move easily to any ocean (and one does not have to contend with complicated visa-application procedures in order to cross an international border).

Although progress in sea-floor research was made possible by the ardent support of scientists, nobody could have predicted that, within a mere 20 years, it would reach such remarkable heights, nor that it would have such a significant impact on the theory of continental drift, which had been revived by the studies in paleomagnetism. Penetrating "insight" must have been at work, even though prediction of the exact future course of research was impossible.

A major contribution to the exploration of the world's oceans was made by the Lamont Geological Observatory of Columbia University, established in 1949. Its first director was the late Maurice Ewing, a man of outstanding leadership and an inquiring mind. The research projects he directed all over the world produced a number of new findings and theories. After more than 20 years of directorship, Ewing moved to the University of Texas in 1972 and died in early 1974, leaving behind important achievements in almost every aspect of ocean-floor studies. The Observatory is now called the Lamont-Doherty Geological Observatory and is still making important contributions under the directorship of Manik Talwani.

The Scripps Institution of Oceanography was established at the turn of the century. Roger Revelle, its director from 1948 to 1964, initiated its emphasis on ocean-floor research. Since the 1950s, the Institution has conducted a number of large-scale expeditions, largely in the Pacific and Indian Oceans and especially in the eastern Pacific.

I have had the privilege of spending considerable time at both these outstanding institutions. It was quite spectacular to observe both research groups making one new discovery after another. The oceanographic institutions of other countries, such as Great Britain, the USSR, Japan, France, Canada, and Germany, have also joined this dynamic effort, and groups of many countries have cooperated in a number of joint projects. For after all, oceans belong to all the nations!

## The Advent of New Techniques

The Japanese research vessel *Hakuho Maru* exemplifies today's modern vessel, fitted with highly advanced survey equipment (Figure 2-1). The most basic piece of equipment required for sea-floor research is a precision sonic depth recorder. In the past, the depth of the

FIGURE 2-1
The research vessel *Hakuho-Maru* (3225 tons). [Courtesy Ocean Research Institute, University of Tokyo.]

ocean was determined by suspending a lead weight from a rope, and then measuring the length of rope needed for the lead to reach the bottom. This procedure required an enormous amount of time and energy. In the 1920s the so-called echo-sounding method was introduced, enabling us to measure the depth of the ocean by sending forth sonic waves from the vessel and recording the length of time they took to echo back. This technique was subsequently highly refined, and by the 1950s it was possible to measure ocean depths throughout the world to a depth of ten thousand meters. The resolving power of the modern precision depth recorder is better than one part in 5000, so that a change as small as one meter in the depth to the ocean bottom can be detected, even if the total depth is as great as 5000 meters. The mapping of the topography of the ocean floor with this device was one of the first developments in marine geology, and it is the first step in any individual project.

Based on the vast amount of data available, the late B. Heezen and M. Tharp of Lamont-Doherty Observatory compiled the now well known birds' eye view of the world's oceans with its spectacular overview of the sea-floor topography. A simplified view of the world ocean topography is shown in Figure 2-2.

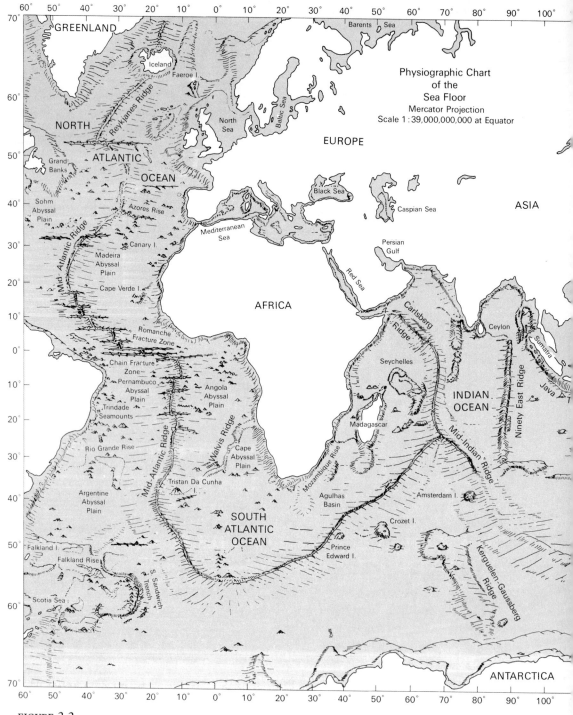

FIGURE 2-2
The floor of the oceans. [Courtesy Hubbard Scientific Company.]

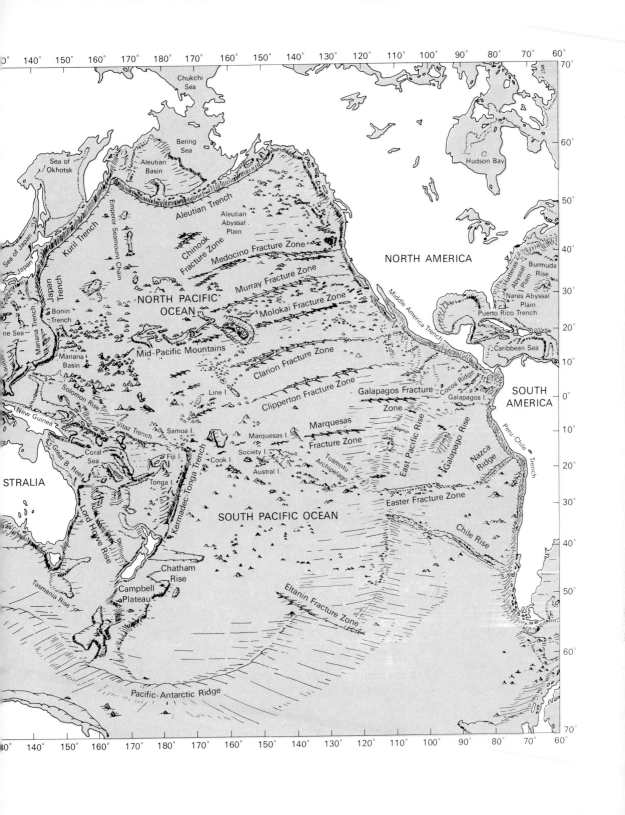

For more sophisticated investigation, seismic waves are used, as on land. It is possible to observe the propagation of seismic waves artificially created by an underwater explosion near the surface of the ocean. This area of study is called marine-explosion seismology. The development of this technique in the 1940s and 1950s is due in great part to American and British scientists. It was this technique that made it possible to determine the suboceanic crustal structure. Recently, another method—the air-gun method—has been developed: a series of waves are generated by shooting an air gun from the ship. The structure of the upper part of the oceanic crust is revealed in remarkable detail by the continuous *reflections* of those waves from buried layers of sediment. Figure 2-3 is an example of the results obtained by this method. The conventional explosion method is now used largely to determine the deeper structure of the ocean crust and upper mantle.

Yet another basic method for determining ocean crustal structure is the measurement of gravity. Since Vening Meinesz's surveys were confined to the limitations of the submarine and consumed much time and energy, the development of a technique of marine-gravity measurement, which could be used on a surface ship, was highly desirable. Thanks to the endeavors of many scientists in various countries, several kinds of extremely intricate surface-ship gravity meters are in use today, producing large quantities of valuable data. C. Tsuboi led the research in Japan, and Y. Tomoda and others successfully developed a ship-borne gravity meter, now installed on the research vessel *Hakuho Maru*.

Measurement of the geomagnetism of ocean areas has become another important research area. The conventional method of geomagnetic field measurement on land had required the precise measurement of the delicate movement of a suspended magnet—a task almost impossible to perform on a rolling vessel. Then in the 1950s a technique for measuring oceanic magnetism, founded on an entirely new concept, was developed. The device is called a nuclear resonance type magnetometer, or *proton-precession* magnetometer. It is known that, in a substance such as water, each proton is constantly spinning like a top and in addition has its own magnetic moment; when the proton is placed in a magnetic field, it undergoes a *precessional* motion like that of a toy top when it is placed in a gravitational field. The working of the magnetometer is based on the fact that the frequency of this proton motion is precisely proportional to the intensity of the magnetic field. In taking measurements it is not necessary

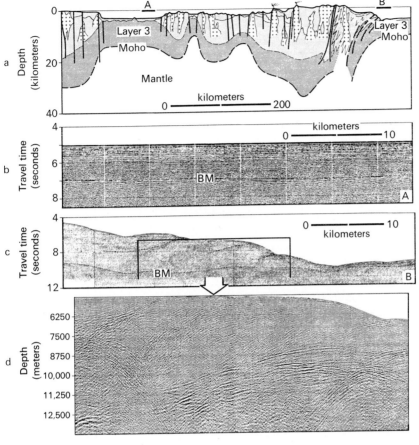

FIGURE 2-3
Example of modern seismic reflection records for the Japan Sea and Japan Trench. (a)
The schematic cross section of the inferred crustal structure, with points A and B
representing the sections at which seismic recordings were taken.
(b) The number of seconds required for a sonic wave to travel at point A in the Japan
Sea. The symbol BM designates the sea-floor basement beneath the sediments.
(c) The seconds required for a sonic wave to travel at point B in the Japan Trench.
(d) The detail of the landward wall of the Japan Trench—designated by the rectangular
outline in part (c)—and its depth in meters. [After R. H. Beck, P. Lehner, et al., "New
Geophysical Data on Key Problems of Global Tectonics." *Proceedings of the Ninth
World Petroleum Congress*, 1975.]

to level or align the magnetic sensor. Consequently it is now a
routine operation to measure the total intensity of the geomagnetic
field from a vessel cruising over the ocean surface. The sensor of the
magnetometer is towed behind the vessel at a distance sufficient to

avoid the effect of the magnetism of the vessel (most vessels being of steel and therefore highly magnetic). Surveys of geomagnetism, conducted with this method, have contributed greatly to research in the science of the ocean floor and to the development of the new view of the earth, as we shall see in this and the following chapters.

Another technique is to measure from a ship the heat that escapes from the earth's interior through the sea floor. Because of its high internal temperature, the earth constantly emanates a certain amount of heat. The quantity of this heat is called *terrestrial heat flow*. The rate of heat flow can be estimated from the measurement of the *geothermal gradient*, which is the rate at which the temperature increases with depth, and the *thermal conductivity* of the strata. The thermal conductivity of a material indicates how effectively heat is transferred by the material. The rate of terrestrial heat flow is obtained simply by multiplying the value of the geothermal gradient by that of the thermal conductivity. The measurement of terrestrial heat flow on land is made by measuring first the geothermal gradient in a mine or a deep well, and then the thermal conductivity of rock samples from the mine or well. Previously, however, measurement at sea posed a practical difficulty, since drilling a hole in the middle of the ocean floor is extremely difficult and expensive. This problem was solved in the 1950s by Bullard and his colleagues, who devised an instrument with a probe a few meters long that contained thermometers mounted along its axis. The probe is lowered until it penetrates the ocean floor and the temperatures at the different points along the probe are measured. From these temperatures the geothermal gradient is easily calculated. On land a deep hole is an absolute necessity for accurate measurement of the geothermal gradient, because the temperature near the surface is affected by the changes of air temperature. Thanks to the enormous thermal inertia of water, the water temperature in the depths of the ocean is constant, so that the necessity of drilling a deep well is eliminated.

Another development that has recently contributed greatly to the progress in marine geoscience is the use of satellites to determine the position of vessels at sea. If one does not know the exact position of the vessel, the measurement of gravity or magnetism means very little, no matter how precise it may be. Usually navigators estimate the position of the ship by taking readings of the sun and stars. Obviously this astronomical method is not applicable in bad weather. But even in fair weather, most estimates are inaccurate by a few miles. In recent years various methods have been developed to determine the position of ships by receiving electromagnetic waves from coastal stations. This method is accurate to within one or two miles under favorable conditions. Yet, as the measurements taken by various surveys became more and more precise, any inaccuracies, however small, came to be considered unsatisfactory. Satellite navigation is a method of determining the position of a ship by receiving the electromagnetic waves transmitted by a satellite. The ship's position can be automatically determined each time the satellite comes around (approximately once an hour) and it is accurate to within 100 meters. A significant improvement over earlier methods, this procedure is now widely used.

All of these new techniques have led to the discovery of important new facts; and the synthesis of these discoveries has molded a new science. Progress in the development of advanced techniques is unending. In the past several years, we have learned how to take various types of measurements by means of instruments set directly on the ocean bottom. *Ocean-bottom seismometers,* in particular, have been successful in providing us with information that is difficult to obtain otherwise. *Ocean-bottom magnetometers,* which will be useful in assessing the thermal state of the upper mantle under oceans, are also being developed. It is hoped that these new techniques will open up yet other new horizons in our science.

## The Ocean Floor

The map in Figure 2-2 shows the topography of the ocean floor. Note for example the long line of topographic highs stretching along the middle of the Atlantic Ocean. This is the great submarine mountain range called the *Mid-Atlantic Ridge.* The existence of such a ridge was already suspected at the end of the 19th century when the first submarine cable was being laid. It was then called the "Telegraph Plateau." A survey undertaken from the famous German vessel, the *Meteor,* from 1925 through 1927, led to the assertion that this "plateau" was a long one, extending practically the entire length of the Atlantic. Since then further studies have revealed that similar mid-oceanic ridges exist in the Pacific and Indian Oceans as the map also shows.

As is also evident in the map, a narrow chain of oceanic deeps lies in the circumferential regions of the Pacific near the island arcs. These deeps are called *trenches.* Between mid-oceanic ridges and continents or trenches, the ocean floor is spacious and quite flat, with numerous seamounts jutting up. If we consider the oceanic ridges and the margins as anomalous areas, the flatter basins between ridges and margins would represent the state of the average or normal ocean floor. The suboceanic crustal structure was first determined by explosion seismology in these basins.

As pointed out in Chapter 1, the oceanic crust in most places is thinner than 10 kilometers and lacks the thick granitic layer invariably found in the continental crust. Recall also that under a thin veneer of sediments, the main layers of the oceanic crust are composed of rocks like the basalts, which have a density and seismic-wave velocity higher than the granitic rocks. The lower layer of oceanic crust probably consists of such rocks as gabbro (the intrusive equivalent of basalts) and serpentinite (hydrated peridotite).

The first heat-flow measurements taken at sea were a surprise. We knew that the heat flowing out of the earth was generated by the decay of the radioactive elements uranium, thorium, and potassium. Chemical analysis had shown that these elements were much more highly concentrated in granitic rocks than in basalt, gabbro, and peridotite. Consequently, scientists expected that average heat flow from the ocean floor would be small, probably no more than one-tenth of the average flow from the continental crust. Those who conducted the early measurements of the heat flow of the ocean floor sought to confirm this point. The first of these measurements was undertaken in 1954 by Bullard and others. The result was unexpected. The heat flow of the average oceanic region was not as small as it was expected to be; in fact, it was about equal to that of the continental region. The similarities in the quantity of heat flow from both continental and oceanic regions, despite the great difference between them in the amount of crustal heat sources, would seem to suggest a greater supply of heat flow from *under* the crust in oceanic regions. The extra heat must be coming from the mantle. This was a significant discovery, as will be seen in the following chapters.

## The Mid-Oceanic Ridges

Among suboceanic topographic structures, the mid-oceanic ridges are the largest (Figure 2-2). The Mid-Atlantic Ridge, for instance, runs from the Arctic Ocean through the middle of the Atlantic, passes the African coast and the Cape of Good Hope, to the Indian

Ocean, from which it continues to the Pacific, practically encircling the entire earth. These mid-oceanic ridges are more than 3000 meters high and more than 2000 kilometers wide. In fact they surpass both the Alps and the Himalayas in scale. It was Ewing and Heezen who first realized that the earth is encircled by such grand-scale sub-oceanic ridges. This important discovery resulted from observations indicating that mid-oceanic ridges could be recognized as such not only because of their singular topography but also because of their seismic activity. Ewing and Heezen were able to predict the existence of ridges, even where a topographic survey had not yet been made, by the linear pattern of the occurrence of earthquakes in oceans. Figure 2-4 shows the world distribution of epicenters—the point on the earth's surface directly above an earthquake focus—demonstrating that most oceanic quakes occur along the mid-oceanic ridges.

In 1953, Heezen and Tharp made yet another important observation: a deep valley winds along the axial line of the Mid-Atlantic Ridge. From the cross-section of the ridge, shown in Figure 2-5, it would appear that this valley was formed by the splitting of the ridge. In contrast to the great mountain ranges on land, which consist mostly of *sedimentary* rock and show evidence of folding created by *compression* from both sides, mid-oceanic ridges are mostly of *volcanic* origin and show features that seem to have been caused by *tension.*

Yet another important discovery about mid-oceanic ridges was that the heat flow was considerably greater at their crests. During 1961 and 1962, I joined R. P. von Herzen of the Scripps Institution of Oceanography to launch a thorough survey of the East Pacific Rise, a feature analogous to the Mid-Atlantic Ridge. Previous surveys of this area had indicated that heat flow might be high at the ridge crest. At that time, in the areas around Japan, the survey of marine terrestrial heat flow was limited to a few sites, although the survey of heat flow on land was quite advanced (Chapter 5). Therefore, the three-month survey on board the research vessel *Spencer F. Baird* in the southeast Pacific, in which more than 300 measurements were taken, was a great revelation to me: today's marine geophysical work must be conducted with this kind of exhaustiveness. The days when conclusions could be drawn from a couple of measurements are over. Some of the results obtained during this cruise are shown in Figure 2-6. Notice the remarkably high terrestrial heat flow at the crest of the East Pacific Rise. We also found that the high heat flow occurs only within two very narrow zones situated on the crest of the East Pacific

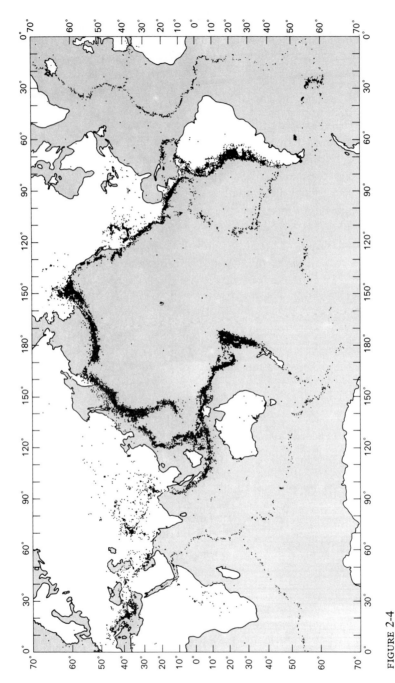

FIGURE 2-4
Diagram showing the distribution of earthquake foci. Notice that those in the oceans are concentrated in the mid-oceanic ridges. [After M. Barazangi and J. Dorman, "World Seismicity Map Compiled from ESSA Coast and Geodetic Survey Epicenter Data, 1961–1967." *Seismol. Soc. Amer. Bull.* **59**, p. 369, 1969.]

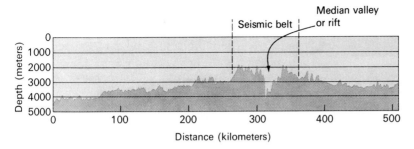

FIGURE 2-5
Profile of the Mid-Atlantic Ridge. [After B. C. Heezen, in S. K. Runcorn, Ed., *Continental Drift*. Academic Press, 1962.]

Rise. Immediately adjacent to these zones, curiously low heat flow values were observed. Such a juxtaposition of high and low heat flow values was later found to be a characteristic of active mid-oceanic ridges in general. This phenomenon is now interpreted as due to the hydrothermal activity in the crust at the crestal part of the ridges, and is under intensive investigation. The mid-oceanic ridges, then, are gigantic topographic highs where heat and tensional forces from the interior of the earth are at work.

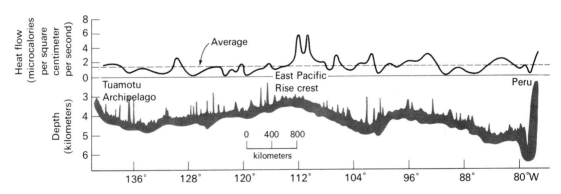

FIGURE 2-6
Profile of heat flow (in microcalories per square centimeter per second) and topography across the East Pacific Rise from the South American coast to Tuamotu archipelago. The horizontal dashed line shows the world average of heat flow. [After R. P. von Herzen and S. Uyeda, "Heat Flow Through the Eastern Pacific Ocean Floor." *J. Geophys. Res.* **68**, p. 4219, 1963. Copyrighted by American Geophysical Union.]

As research progressed in the Atlantic and the Pacific, interest in the mid-oceanic ridges of the Indian Ocean also grew. The International Indian Ocean Expedition Project (1959–1965) materialized out of the cooperative spirit created by the International Geophysical Year (1957–1958) and the Upper Mantle Project (1962–1970). The United States, England, the USSR, France, Germany, and Japan all sent out scientists and research vessels to undertake extensive research.

The Indian Ocean floor is also shown in Figure 2-2. There too the system of mid-oceanic ridges is extensive. It is also quite complex. For example, the extension of the Mid-Atlantic Ridge circles the southern tip of Africa into the Indian Ocean, where it bifurcates. One of the branches extends eastward to the south of Australia and eventually reaches the Pacific Ocean. The other branch passes through the Gulf of Aden and enters the Red Sea. If we assume that something hot is welling up from beneath the mid-oceanic ridges and rending them apart, then it is logical to assume that the Gulf of Aden and the Red Sea currently constitute the now active rift that is causing the two continents to split (see Figure 2-7). In fact, we now have ample evidence showing that this inference is indeed true. As we will see later, the data collected by the International Indian Ocean Expedition and other expeditions proved to be extremely valuable, once the "new view of the earth" had shown us how to interpret these data in terms of the evolution of the world's ocean.

### The Oceanic Trenches

It might be logical to assume the deepest part of the ocean is in the middle of the ocean. But the actuality is quite the contrary. As the topographic map in Figure 2-2 shows, the deepest waters lie close to the land, in the margin of the ocean. The middle of the ocean is shallower, owing to the presence of the mid-oceanic ridges. This distribution is similar to that of the highest mountains on the continents. With a few exceptions, such as the Himalayas, most of the high mountain ranges are not in the middle of the continents but at their margins, facing the deep oceanic trenches.

The seemingly paradoxical distribution of ocean depths and shallows, and of mountain ranges, is closely related to the origin of the continents and the ocean floor. The trenches are highly developed in such areas as the western margin of the Pacific Ocean—from Alaska to the Aleutians, the Kuriles, Japan, Izu-Bonin and the Marianas, and Tonga-Kermadec—and the southeastern margin of the Pacific Ocean, the west coast of South America. Along the continental side of the oceanic trenches lie island arcs and continental arcs where the

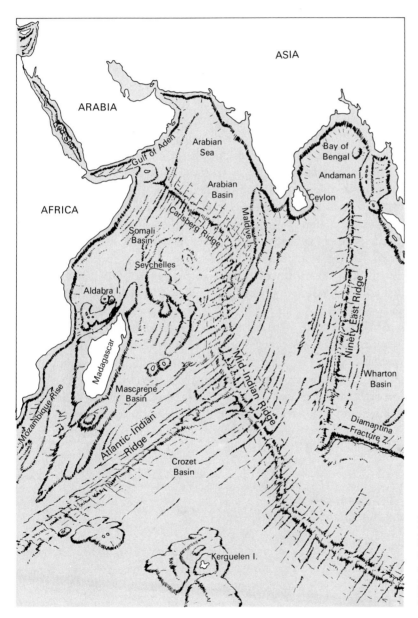

FIGURE 2-7
Portion of the sea floor, showing at the upper left the active rift between Asia and Africa, through the Gulf of Aden and the Red Sea. [Courtesy Hubbard Scientific Company.]

seismic and volcanic activity is vigorous. Since oceanic trenches and arcs always occur together, the two need to be considered as a pair. A system of island arcs and trenches is well developed in the Indonesian region also.

The pioneers of research on island arcs were the Dutch scientists who conducted investigations on the Indonesian archipelago as early as the 1920s. The measurement of gravity in the ocean by Meinesz, mentioned earlier, was one such pioneering investigation. The results of his measurements showed the gravity to be unusually low in the oceanic trenches. Since the trenches are filled with water (density of 1.0 grams per cubic centimeter) instead of rock (density of 2.6 to 3.0 grams per cubic centimeter), at first this finding appears only logical. And yet, we must not forget the principle of isostasy (see page 17) according to which material beneath the mountains must be lighter in order to maintain the buoyancy necessary to support them. By the same reasoning, heavy material must exist under the ocean trenches in order to maintain the depression. However, the low gravity readings in the trenches showed that the heavy material required for isostatic balance was definitely *not* present. Meinesz's discovery produced the following important question: what is it that holds down the oceanic trenches? Unless something were holding the trench floor down, it would be expected to rise and soon disappear, just as a mountain range not supported by a light root would sink and soon disappear.

Another feature of the oceanic trenches is their low heat flow, in contrast to the unusually high heat flow along the crests of submarine ridges. This characteristic suggests the following about the origin of the oceanic ridges and oceanic trenches: within the mantle there is a flow of material that ascends at the oceanic ridges and sinks at the oceanic trenches. Additional details of these geophysical measurements in the trench-arc areas and their geotectonic significance will be discussed in Chapter 5.

## The Geomagnetism of
## the Sea Floor: A Riddle

The new methods of measuring geomagnetism with ship-borne equipment (see page 48) spearheaded the research in deep-sea geophysics, yielding remarkable results on the distribution of geomagnetic anomalies in the ocean. Very active geomagnetic surveys in the East Pacific area were initiated by the Scripps Institution of Oceanography and U.S. Coast and Geodetic Survey and by such scientists as V. Vacquier, R. Mason, and A. Raff.

Figure 2-8 is a schematic representation of the distinct striped pattern of geomagnetic anomalies in the East Pacific extending north and south—a phenomenon never seen on the land. Although the earth's magnetic field forms a dipole pattern (see Figure 1-11), actual measurements reveal some deviations. These deviations are called *geomagnetic anomalies*. There are two general types of geomagnetic anomalies: (1) the large-scale anomalies with dimensions of thousands of kilometers, and (2) the more local ones. Large-scale anomalies are called *regional anomalies* and are believed to be caused by dynamo actions in the earth's core. The *local anomalies* are caused by the inhomogeneous magnetization of the crustal materials, and the anomalies of the ocean floor are definitely of this type. The striped pattern of the geomagnetic anomalies shown in Figure 2-8 therefore suggests that the oceanic crust is magnetized in stripes. When this remarkable striped pattern was discovered, the cause of it became an important riddle in marine geophysics.

There is another important feature shown in Figure 2-8 that is surprising. The stripes appear to be severed in several places. Closer examination of these lines of severance has indicated a displacement of the stripes of more than 100 kilometers; in a few places (not shown in the figure) displacement is so extreme that it is more than 1000 kilometers. These lines of severance are coincident with topographical discontinuities known as *fracture zones*. This displacement of magnetic stripes along fracture zones was a key observation in the theory of plate tectonics.

One of the reasons that many scientists rejected Wegener's ideas was that they could not accept the idea of the continents having moved thousands of kilometers. Yet it seemed clear that the distances that adjacent parts of the ocean floor moved from one another, as estimated by the displacement of geomagnetic stripes, were more than 1000 kilometers.* This indicated a more mobilist view of earth

--------

*The American Miscellaneous Society, which makes annual awards to people for extraordinary accomplishments, presented Victor Vacquier—a scientist who contributed to the monumental survey of geomagnetism in the East Pacific—their 1966 award, citing him as "The man who moved the ocean floor several thousand kilometers." Vacquier was visiting the University of Tokyo at that time and, as I remember it, the award was a stuffed albatross. This incident was evidence of oceanographers' interest in the displacement of the striped pattern in the marine geomagnetic anomalies.

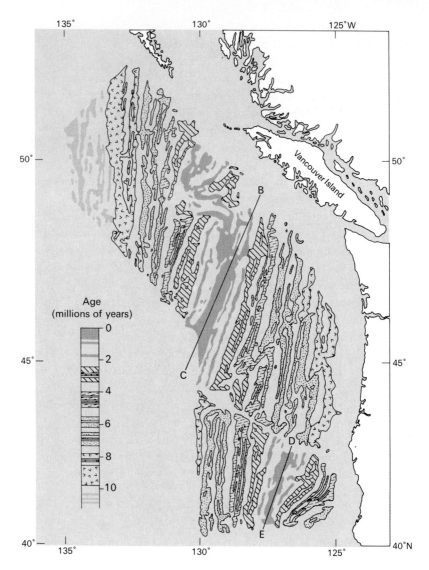

FIGURE 2-8
Summary diagram of total-field magnetic anomalies southwest of Vancouver
Island. Areas of positive anomalies are shaded and are thought to approximate the
areas of normal magnetization in the oceanic crust. The different patterns
represent the different geological ages as shown on the vertical scale. The central
anomalies (solid gray) coincide with the ridge crests—the Juan de Fuca to the
north (BC) and the Gorda to the south (DE). Note that BC and DE are offset by a
fracture zone CD. [From F. J. Vine, "Magnetic Anomalies Associated with
Mid-Ocean Ridges," in R. H. Phinney, Ed., *The History of the Earth's Crust*.
Copyright © 1968 by Princeton University Press. Redrawn by permission of
Princeton University Press.]

history, but the question about the origin of the striped pattern itself remained. When it was finally answered, it was found that the displacement of magnetic stripes was due *not* to ordinary fault movement, but to something even more interesting. This revelation suggested a whole new concept of the earth—a topic we will discuss further in the next chapter.

## Chapter 3

# The Hypothesis of the Spreading Ocean Floor: A Synthesis

### Geopoetry

As the theory of continental drift regained momentum and as progress in deep-sea geophysics directed scientists' attention toward the significance of the mid-oceanic ridges, Holmes' suggestion that the continents are carried by convection within the mantle (Chapter 1) began to seem more interesting. It was at this time that the late H. Hess of Princeton published a daring paper that attempted to establish a new concept of the earth, discarding such fixed ideas as an immovable earth or an unchanging ocean. The paper, entitled "History of Ocean Basins," was widely circulated among scholars before its publication, so that by the time it was published in 1962 its hypothesis was already widely known. In his introduction Hess stated, "I shall consider this paper an essay of geopoetry." The earth as depicted by this geopoetry is schematically represented in Figure 3-1. The mid-oceanic ridges are outlets for the substances welling up from the mantle; in other words, they are regions in which the conveyor belt, originally proposed by Holmes (page 38), is exposed on the surface. It is precisely there that the new suboceanic crust is born. This new ocean floor spreads out on both sides of the oceanic ridges, and descends again into the mantle at the oceanic trenches.

The speed of the flow of the "conveyor belt" is considered to be several centimeters per year. This means that it takes no more than about two hundred million years for the ocean floor, welling up at the

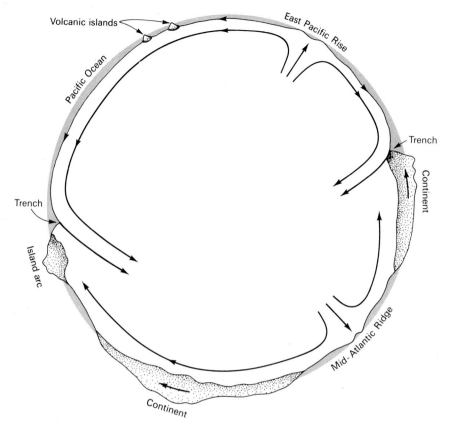

Volcanic islands

East Pacific Rise

Pacific Ocean

Trench

Continent

Trench

Island arc

Mid-Atlantic Ridge

Continent

FIGURE 3-1
Schematic cross section of the earth based on the sea-floor spreading hypothesis.

mid-oceanic ridges, to move across the ocean, and sink into the oceanic trenches. It would thus appear that the ocean floor is not permanent, but is constantly being renewed. Continents, however, cannot sink as readily into the earth's interior on the conveyor belt because they are much too light. Therefore they are semipermanent. This idea explains the two riddles that had haunted marine geologists for decades: (1) Why have rocks older than 150 million years never been found on the ocean floor? (2) Why are the sediments of the ocean floor so thin? The ocean itself is several thousand millions of years old, but its floor has been continually changing!

Hess, in his paper, stressed one of his original ideas, that the oceanic crust is probably composed of serpentinized peridotite. The upper mantle is considered to be composed mainly of peridotite that contains water. It has been proved experimentally that, under high temperatures, peridotite and water are separate, but at temperatures below about 500°C, peridotite reacts with water and becomes serpentinite. Hess maintained that peridotite, ascending from the depth of the mantle, is serpentinized as it nears the surface and forms the oceanic crust at mid-oceanic ridges. When it descends into the mantle at the oceanic trenches, as it is reheated to above 500°C, water is released. Hess considered this released water to be the source of sea water.

This remarkable and now widely accepted theory is known as "sea-floor spreading," but the true originator of the idea was for a while the subject of some controversy. Shortly before the publication of the Hess paper on geopoetry, another well known American scientist R. Dietz (1961) published a similar hypothesis. Although it was Dietz who coined the intriguing term "sea-floor spreading," some debate on which man first came up with the hypothesis ensued, because the Hess paper had been read extensively before its publication. Later Dietz himself freely acknowledged Hess's priority. To me, Dietz's paper was just as enlightening as Hess's; and, in some respects, Dietz provided a clearer explanation of the hypothesis than Hess did, though perhaps with less geopoetry.

In fact, in retrospect, it seems to me that quite a number of scientists were nursing similar ideas around 1961 and 1962. This fact makes the controversy seem less important and perhaps its resolution should be left to the professional historians of science. I personally would like to see more credit go to Holmes who proposed his conveyor-belt hypothesis thirty years earlier.

Dietz, in his paper, supported the prevailing assumption that the oceanic crust is composed of basaltic gabbro, but departed from the accepted view in his assertion that the mantle is composed not of peridotite but of eclogite, which is formed from gabbro at very high pressures. Assumptions like this are made in an attempt to answer very basic questions: What is crust? What is mantle? What is the nature of the Mohorovičić discontinuity, or Moho? Opinions are still divided on these questions.

The more common view is, as explained in Chapter 1, that the Moho is the boundary between two layers of different chemical compositions, gabbro and peridotite, gabbro constituting the lower crust and peridotite the upper mantle. Another school of thought, however, holds that it is the boundary *not* between different materials, but between different *states* of the same material. It is interesting to observe that, though their models were different, both Hess and Dietz regarded the Moho as a boundary between states rather than between materials. This common point of view might have been a mere coincidence, but certainly it reminds us of our inability to answer such a basic question, "What is the Moho, which lies just a few kilometers below the sea bottom?"

Dietz attached little significance to the Mohorovičić discontinuity in the context of sea-floor spreading. He chose to call the earth's surface layer, to a depth of about 70 kilometers, the *lithosphere*, a term that had been used by earlier geologists to refer to the earth's outer layer of solid rock. Dietz regarded the lithosphere as a dynamic unit that moved as a single entity. He also maintained that underneath this layer was a slightly softer layer, the *asthenosphere*, which allowed the lithosphere to move. This argument, put forward in 1962, contained a foreshadow of the concept of plate tectonics that emerged five years later.

## It Is the Sea Floor That Moves

J. Tuzo Wilson of Canada was an enthusiastic supporter of the hypothesis of sea-floor spreading. Wilson had espoused many original and startling ideas in the past, at one time supporting the theory of a contracting earth and later supporting that of an expanding earth. In the early 1960s, he began to maintain that the theory of convection within the mantle would explain many phenomena, including that of continental drift. Scholars who change their opinions too often usually lose the respect of their colleagues, but Wilson's insight and originality appear to have made him an exception. He seems to operate on the principle that everything should start as a hypothesis; once adopting a hypothesis, he pursues its consequences so thoroughly that finally some means of testing it emerges. If it stands the test, he pursues it further. If it is disproven by this process, he discards it. At this stage, usually another hypothesis has emerged to take its place.

Wilson approached the problem in this way. He proposed that— given that the Atlantic Ocean was a gigantic rift and that centers of volcanic activity were localized at or near the center of the rift—the ages of the islands scattered throughout the Atlantic, all volcanic in origin, should increase the further they had migrated from the Mid-Atlantic Ridge. Having gathered and assessed all the data available at the time (in the early 1960s), he concluded that this assumption was in fact correct. As Figure 3-2 shows, the longer the distance of an island from the Mid-Atlantic Ridge, the greater the age. For example, the island of Ascension, close to the Mid-Atlantic Ridge, is no more

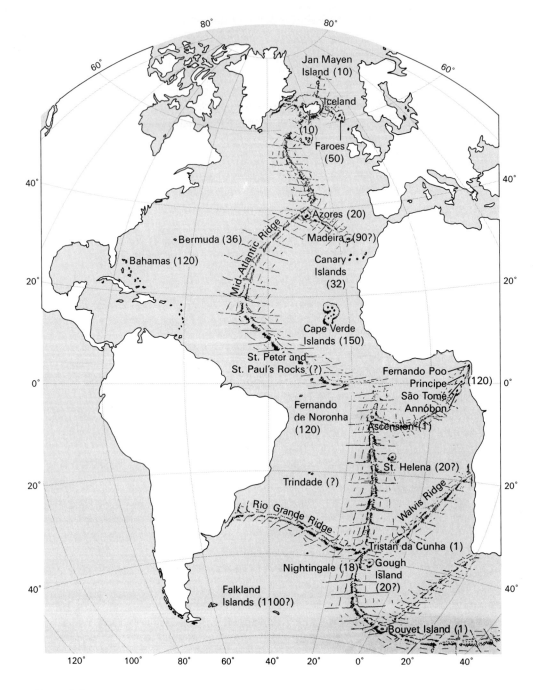

FIGURE 3-2
Ages of the Atlantic islands, as indicated by the ages of the oldest rocks found on them. The numbers in parentheses give the ages in millions of years. [After J. T. Wilson, "Continental Drift." Copyright © 1963 by Scientific American, Inc. All rights reserved.]

than a million years old, St. Helena, further from the same ridge is thought to be around 20 million years old, and the islands near the west coast of the African continent, such as Fernando Poo and Principe, are 120 million years old. This pattern coincides beautifully with the assumption, based on the theory of continental drift, that the Atlantic began to form during the Jurassic period about 200 million years ago. A similar examination of the chains of volcanic islands in the Pacific (see Figure 2-2) reveals that they too conform to this pattern, having gradually migrated from their points of origin. For example, the islands of the Hawaiian Archipelago, which line up northwest of the Island of Hawaii, have been shown to increase in age northwestward. The magma source for this archipelago is believed to be not a ridge but the point at which the Island of Hawaii is presently located. This source is an example of a "hot spot," which will be described in Chapter 5.

As these and other pieces of evidence accumulated, the concept of "the moving sea floor" gradually gained general acceptance. Wilson once pointed out, "If indeed the Earth is, in its own slow way, a very dynamic body, and we have regarded it as essentially static, we need to discard most of our old theories and books and start again with a new viewpoint and a new science."[*]

## The Riddle of the Striped Patterns

One of the significant events in marine geophysics during the late 1950s was the discovery of the striped patterns of geomagnetic anomalies on the ocean floor and their offsets, discussed in Chapter 2. Why these spectacular striped patterns existed at all, however, was still unknown. The anomalies are 20 to 30 kilometers wide and hundreds of kilometers long with an amplitude of several hundred gammas.[†] We have seen that the source or cause of a magnetic anomaly may be either a current flowing in the earth's core or a body of magnetic rock within some 50 kilometers of the earth's surface. In order for the striped anomalies to be produced, the sources would have to lie directly beneath the striped belts—probably at a shallow depth; if the source were deep, the surface anomalies would not be as distinct as observed. Thus it seemed likely that the source consisted

[*]J. Tuzo Wilson, "A Reply to V. V. Beloussov," *Geotimes*, December 1968, p. 22.

[†]The gamma is a unit used in measuring the geomagnetic fields. The geomagnetic field in Tokyo is about 46,000 gammas and that in New York about 57,000 gammas.

of long, linear bodies of magnetic rock. One possibility was that the positive anomalies, where the field was unusually strong, were underlaid by prisms of strongly magnetized intrusive or volcanic rocks, whereas the negative anomalies, where the field was weak, were underlaid by steep-walled valleys filled with sediments, which are only weakly magnetic. Yet no one was really able to explain the origin and nature of the rocks that had caused the stripes, or even able to prove that such sources existed.

In 1963, two young British scientists, F. Vine and D. Matthews came up with an enlightening explanation. They proposed that the pattern of magnetization in the crust that causes the anomalies to be striped is due not to variations in the *intensity* of magnetization, but to changes in the *direction* of that magnetization. Beneath the positive anomalies the rocks are *normally* magnetized in a direction parallel to the present field. Beneath the negative anomalies the rocks are *reversely* magnetized in the opposite direction. They further maintained that this phenomenon calls for *no* new hypothesis. According to Vine and Matthews, the striped pattern of geomagnetic anomalies is a logical consequence of the combination of two fundamentally important but independently established phenomena—first, the spreading of the ocean floor that wells up at the crest of the mid-oceanic ridges and, second, the changes in polarity of the geomagnetic field, which reverses every several hundred thousand years or so (see the discussion on geomagnetism, page 30). Let us examine this hypothesis in further detail.

According to the sea-floor spreading hypothesis, magma from the hot mantle, as it ascends to the oceanic ridge to form the new ocean floor, cools through its Curie point. This is the moment at which the newborn crust is magnetized in the direction—either normal or reversed—of the geomagnetic field prevailing at that particular period or epoch. As the ocean floor slowly spreads away from the oceanic ridge, it is inevitable that a strip of ocean floor formed during a normal polarity epoch will be adjacent to a strip in which the magnetization is reversed, producing in a striped pattern. The spreading ocean floor—likened by Holmes to a conveyer belt—can now be described as a type of tape recorder as well. The oceanic crust ascending from the mantle is the magnetic tape that records the history of geomagnetic reversals.

L. Morley of Canada published an identical but independently developed explanation for the striped pattern of magnetic anomalies at almost the same time as Vine and Matthews. It is now fairly well known among earth scientists that Morley's paper, which had been submitted earlier to leading British and American scientific journals, was rejected at the time as being too speculative. It is a sad story. In all fairness, perhaps the hypothesis should be called the Vine-Matthews-Morley hypothesis.

## The Ocean Floor as a Tape Recorder

The idea of a spreading ocean floor as proposed by Hess, Dietz, Wilson, and others, although it appealed to many scientists, was originally regarded with some suspicion. The reason was probably that so simple an explanation of such a complex set of natural phenomena tends to invite skepticism. It seems that there are always people who unconsciously think (and perhaps even hope) that nature is too complicated to be explained by a simple idea. The Vine-Matthews-Morley hypothesis—an attempt to explain the striped pattern of magnetic anomalies on the basis of a combination of two hypotheses that were themselves questionable—was also shunned at first by many scientists. The depressing fate of Morley's paper shows how widespread this rejection was. However, the skepticism was overcome in 1966 when a great many findings in support of these new concepts were reported that same year.

Vine reported that he and Wilson had succeeded in interpreting the magnetic anomaly lineations within the framework of the tape-recorder model—not only qualitatively, as Vine and Matthews had done, but also quantitatively. Vine assumed that the speed with which the ocean floor spreads from a given ridge remains constant. Given this constant rate, the width of each magnetic stripe on the ocean floor should be proportional to the length of the duration of each normal and reversed polarity epoch. This quantitative relationship between the known reversal time scale and the width of the magnetic stripes convinced a number of scientists that the idea of sea-floor spreading did indeed have some validity.

How did Vine arrive at his remarkable conclusions? For one thing, he had some very useful and reliable data to work with. A. Cox, R. Doell, and B. Dalrymple had by 1966 worked out the history of geomagnetic field reversals for the past four million years. Using the potassium-argon method,* they had dated magnetic rocks from all

---

*This method of determining the absolute age of rocks is based on the following. The radioactive potassium isotope, $K^{40}$, contained in rocks, disintegrates into the argon isotope, $A^{40}$; it has a half life of $1.42 \times 10^8$ years. Therefore, if the amounts of $K^{40}$ and $A^{40}$ now existing in a rock are determined, the date when $A^{40}$ began to accumulate can be inferred. This date is considered to be the age of the rock.

over the world and had determined exactly how many millions of years ago different geomagnetic field reversals had occurred. Anyone who was a regular visitor to the laboratory of Cox and his colleagues during that period would find that the history of geomagnetism was continually being unraveled, for something new invariably awaited each visit. Figure 3-3a shows the chronology of these geomagnetic field reversals as they were determined from the paleomagnetic research on volcanic rocks on land. In this figure, the center indicates the present time, and the dates on either side, from the most recent to the oldest, are given in units of a million years. The periods during which the polarity was fixed were called *epochs* by Cox and his colleagues, and each epoch was named after one of the pioneers of geomagnetism. The present epoch is called the Brunhes normal epoch, after the French scientist who, as early as 1906, postulated the possibility of the reversal of the geomagnetic field in the past. The preceding epoch, ending about 700 thousand years ago, is called the Matuyama reversed epoch after the Japanese scientist M. Matuyama, mentioned in Chapter 1. Cox and his colleagues also found that, superimposed on the normal and reversed epochs, there were short intervals of less than about 100 thousand years, during which the polarity was opposite from that of the epoch in which they occurred. These intervals are called *events* and are named after the localities in which the evidence of their existence was first discovered. For instance, one of the events is named Olduvai—after the gorge in Africa so well known as the site of spectacular anthropological findings.

Vine undertook the task of comparing the observed profiles of striped magnetic anomalies with the reversal time scale. Assuming reasonable rates for ocean-floor spreading, he came up with the excellent agreement demonstrated by the two curves in Figure 3-3b. The upper curve represents the profile of geomagnetic anomalies that has been theoretically calculated from the model sea floor as shown in Figure 3-3c. This model sea floor was constructed on the basis of (1) the history of geomagnetic field reversals, as demonstrated by Cox, Dalrymple, and Doell (1967), and (2) the theory of sea-floor spreading with a reasonable speed assumed for the rate of spreading. The lower curve is the observed anomaly profile of the East Pacific Rise. It is hard to imagine that the agreement between the computed and the observed profiles is accidental. Similar agreement has been demonstrated for other oceanic ridges such as the Reykjanes Ridge south of Iceland, the Mid-Indian Ridge, and some Antarctic ridges. Although the degree of agreement varies from ridge to ridge, it is usually possible to identify all of the longer epochs. Knowledge of the precise chronology of the geomagnetic field reversals has enabled us to de-

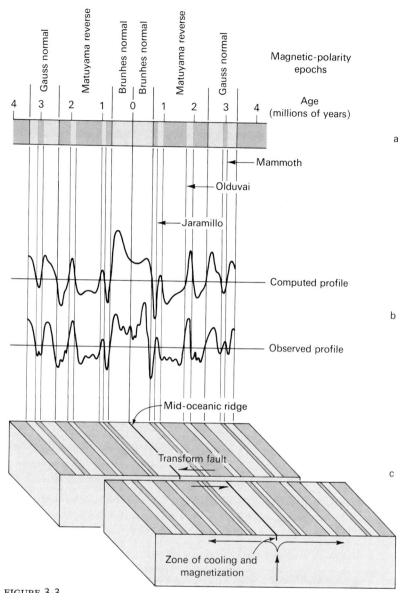

FIGURE 3-3

(a) The geomagnetic polarity epochs and events. Ages are given in millions of years. [After A. Cox, B. Dalrymple, and R. Doell, "Reversals of the Earth's Magnetic Field." Copyright © 1967 by Scientific American, Inc. All rights reserved.]

(b) Comparison of the observed geomagnetic anomaly profile Eltanin-19 (lower curve) with the computed profile (upper curve) for the East Pacific Rise. [After F. J. Vine, "Spreading of the Ocean Floor; New Evidence." *Science* **154** p. 1405. Copyright 1966 by the American Association for the Advancement of Science.]

(c) Model sea floor with the magnetic anomaly stripes produced by the Vine-Matthews-Morley mechanism. A fracture zone, represented by the displacement zone of the two blocks, cuts the ridge and the magnetic stripes, producing a transform fault (see page 74).

termine the absolute rate of the speed of sea-floor spreading. According to Vine and Wilson, the spreading rate is 4 to 5 centimeters per year for the East Pacific Rise, and about 1 centimeter per year for the Reykjanes Ridge. At the end of 1966 Vine published a masterful paper that left little room for doubt about the origin of the magnetic stripes and the rate of sea-floor spreading. Subsequently Vine was asked at a symposium whether he had checked his results statistically. "I never touch statistics," he answered, displaying his total confidence in his work, "I just deal with the facts."

By the end of 1966, J. Heirtzler, W. Pitman, and their colleagues at Lamont-Doherty Geological Observatory had accepted the interpretation of Vine, Matthews, and Wilson, and were beginning to use the width and spacing of the magnetic stripes to determine the rate of sea-floor spreading. In fact, the Lamont-Doherty group and Vine worked on the same anomaly profile across the East Pacific Rise taken by the National Science Foundation ship *Eltanin*—the now classical *Eltanin*-19 profile shown in Figure 3-3b—and both groups arrived at the same velocity of sea-floor spreading. The use of magnetic stripes to measure spreading velocities quickly transformed a speculative idea unworthy of publication into an established technique.

Yet another remarkable report was made in the same year. It was on the research on the remanent magnetism found in sea-floor sediments. Remanent magnetism exists within the sedimentary strata of the ocean floor. As fine particles of magnetic minerals—minute magnets—are deposited on the ocean floor, they tend to settle down, with their magnetization directions aligned with the prevailing geomagnetic field. This gives rise to the remanent magnetism of sediments. However, measurement of the direction of magnetization of ocean sediments is difficult because the magnetization is very weak. Also the sediments are soft and difficult to handle. Scientists at Cambridge University and the Scripps Institution of Oceanography had been attempting to take these measurements since 1957, but without much success. Finally, in 1966, N. Opdyke, B. Glass, and others of the Lamont-Doherty Geological Observatory succeeded in measuring the remanent magnetism in samples from the Antarctic and the North Pacific Oceans with spectacular results. Since sedimentation on the ocean floor is an extremely slow process, a continuous history of geomagnetism during the past several million years can be obtained by measuring the magnetization of sedimentary material only some 10 meters below the sea bottom. The sample material for the measurement, or core sample as it is called, is taken

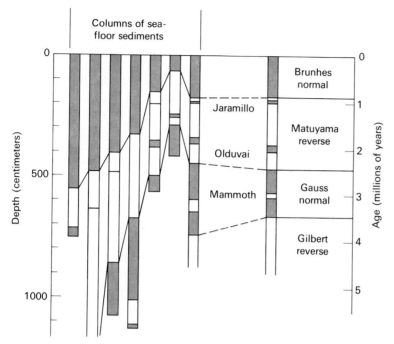

FIGURE 3-4

Normal and reverse magnetization of ocean sediments. The shaded areas in the columns represent normal magnetization; the white areas represent reverse magnetization. [After N. D. Opdyke, B. Glass, J. D. Hayes, and J. Foster, "Paleomagnetic Study of Antarctic Deep Sea Cores." *Science* **154**, p. 349. Copyright 1966 by the American Association for the Advancement of Science.]

from the ocean floor in the form of a vertical column by a pipe, called a corer, which is driven into the sediment. The results from such samples (diagrammed in Figure 3-4) distinctly demonstrate the alternation of normal and reversed magnetization. The major epochs are revealed by these measurements and even some of the events, but the events are not recorded consistently because the sediments are sometimes disturbed by burrowing organisms.

These results have justified the development of a new field of geology known as paleomagnetic stratigraphy. Precise identification of strata can now be accomplished by the examination of the direction of remanent magnetism, another valuable technique for dating rocks. Thus the history of geomagnetic field reversals was established quantitatively through three independent phenomena—the remanent magnetism of volcanic rocks from all of the continents, the magnetic stripes of the ocean floor, which are several tens of kilometers in width and several hundred kilometers in length, and the feeble remanent magnetism of ocean sediments collected from a stratigraphic thickness of 10 meters or less. The remarkable agreement of the results obtained from these three lines of investigation has established conclusively that the ocean floor is a high-fidelity magnetic tape recorder.

## The Transform Fault—
## A Concept of Great Originality

As mentioned toward the end of Chapter 2, Vacquier and his colleagues discovered spectacular offsets, or displacements, in the geomagnetic stripes off the west coast of North America. These offsets profoundly impressed other scientists because they appeared to signify that large-scale displacements between adjacent segments of the ocean floor—some as large as 1000 kilometers—had taken place. Keeping in mind the cause of the striped pattern as explained by Morley, Vine, and Matthews, let us now examine the reason for the offsets of the magnetic stripes. Do these offsets really represent the slippage of adjacent segments of the oceanic crust?

It was J. Tuzo Wilson (1965) who proposed the following interpretation. He suggested that these offsets were not ordinary "transcurrent" faults but an entirely new type of fault, which he called a "transform fault." Figure 3-5 illustrates the difference between the two. In a transcurrent fault (a), the blocks on either side of the fault move in the directions of the arrows, thus causing the whole structure—the mountain range in Figure 3-5a, for example—on either side of the fault to be displaced. This type of fault is widely observed on land and had been the conventional explanation of the observed displacement of magnetic stripes and of oceanic ridges. In the transform fault (Figure 3-5b) the displacement bb' of the mid-oceanic ridge, ab and b'c, does not appear to differ from the displacement caused by the transcurrent fault. But in the sea-floor spreading hypothesis, the mid-oceanic ridges are viewed as dynamic rather than static because the new ocean floor is constantly flowing forth and spreading in the direction of the arrows. If we accept this concept, then it becomes apparent that the transform fault is quite different from the ordinary transcurrent fault. First of all, in the transcurrent fault displacement BB' will increase with time as the faulting activity continues; however, if we consider the mid-oceanic ridge sections ab and b'c to be producing sea floor with equal speed, the displacement

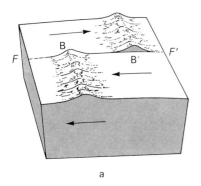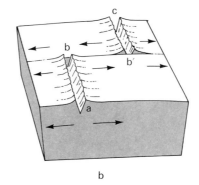

a                                              b

FIGURE 3-5
Two types of faults: (a) transcurrent fault; (b) transform fault.

bb' will not change at all. Moreover, the displacement between the
blocks on either side of the fault occurs only along the portion bb'; in
the portions outside bb' no displacement occurs across the fault. This
has an important implication for seismology. If earthquakes are
caused by displacement between blocks on opposite sides of a fault,
they will occur along the entire length, FF', of the fault in the transcurrent fault (Figure 3-5a), but only along the length bb' between the
ridges in the transform fault (Figure 3-5b). Furthermore, whereas the
*apparent* displacement of the ridge axis is in the same direction in
each type, the actual direction of movement across the fault is different. If it is a transcurrent fault, and you are standing on the south side
of the fault, the earth on the opposite side is moving to your *right*. If it
is a transform fault, however, and you are standing on the south side
of the fault between the ridges, the earth on the opposite side is
moving to your *left*. Wilson realized that the faults that commonly
sever mid-oceanic ridges would have to be of the transform type (see
Figure 3-3c). Figure 3-5 illustrates a fault only between one oceanic
ridge and another, but faults can also exist between an oceanic ridge
and a trench, or between two trenches.

Wilson's significant insight about transform faults occurred because he found it very hard to believe that a large-scale fault causing
an apparent displacement of a thousand kilometers could be a transcurrent fault. Indeed, he found it inconceivable because there was no
explanation for the disappearance, at the ends of the fault, of the
crustal rocks on either side of it. Where do the displaced portions go?
They can't simply vanish without violating the law of conservation of
matter. Wilson found that fracture zones, which had long been

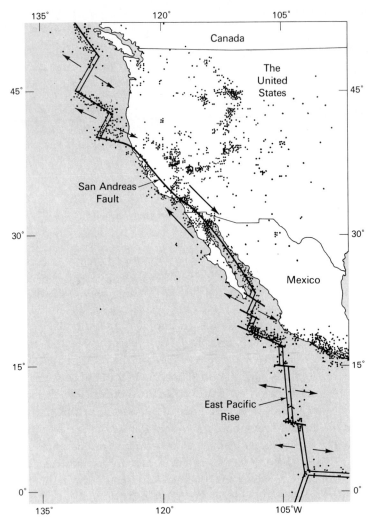

FIGURE 3-6
The San Andreas Fault as a transform fault. The double line represents the spreading ridge; the thick line, the transform fault. The shorter arrows on opposite sides of the East Pacific Rise show the movement of blocks on opposite sides of the rise *away* from each other. Along the San Andreas Fault the long arrows show the movement of the two sides of the fault past each other. The dots designate the earthquake epicenters.

known to oceanographers, provided the key he needed to answer these questions. Between offset ridges, the fracture zones mark the trace of still active transform faults. Away from the offset ridges the fracture zones are no longer active but are scars marking the trace of former transform faults.

The concept of transform faults suggested the possibility that the famous San Andreas Fault (Figure 3-6) might be such a fault. The occurrence of earthquakes on the East Pacific Rise suggested to Wilson that this was a mid-oceanic ridge from which the sea floor was diverging. The East Pacific Rise extends into the south end of the Gulf of California, and the San Andreas Fault emerges from the north end of it. The San Andreas Fault is known for the presence of shearing motion parallel to the fault rather than tensile motion diverging away from the fault. Why should the tensile forces in the East Pacific Rise suddenly change into the horizontal shearing forces of the San Andreas Fault? This question had long been an enigma to scientists. If the San Andreas Fault were considered a transform fault, however, Wilson showed that this enigma could easily be explained as demonstrated in Figure 3-6. To complete his explanation, Wilson needed another mid-oceanic ridge at the north end of the San Andreas transform fault. To prove the actual existence of such a fault, Vine and Wilson began scrutinizing the pattern of geomagnetic anomalies off Vancouver Island. One can imagine their elation when they discovered the symmetry in the geomagnetic stripes between the axes of the two shown in Figure 2-8. Profiles of geomagnetic anomalies across these axes showed an almost perfect agreement with the history of geomagnetic reversals, based on the assumption that the speed of sea-floor spreading is 2.9 centimeters per year. The topographic high along BC in Figure 2-8 was named the Juan de Fuca Ridge, and that along DE the Gorda Ridge.

Wilson once told me that he hit upon the idea of the transform fault while he was cutting out a model of a spreading mid-oceanic ridge from a piece of paper. It should be noted here that this idea occurred to him only after he had carefully examined an enormous amount of data that had already been compiled. Thus although it was his acute insight that brought the immense past efforts of other scientists to fruition, the accomplishment of Vacquier and his colleagues, who produced accurate data on geomagnetic anomalies, must not be overlooked. Their data proved to be most valuable for checking the validity of many theoretical inferences.

## Verification of the Transform Fault

The direction of faulting can be estimated from the analysis of the initial motion of earthquake waves generated by the faulting. Earthquakes occur when a fracture takes place within the earth's interior,

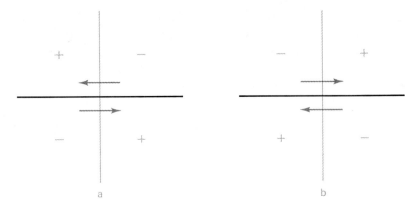

FIGURE 3-7
Radiation patterns for initial *P*-wave motions. The faults are shown by the thick lines, and the arrows indicate the direction of faulting. In regions marked (+), the initial motion is away from the source; in regions marked (−), it is toward the source.

which causes elastic waves to be generated. By analyzing the direction of the initial motion of an earthquake wave, with the aid of seismometers placed around the epicenter, seismologists can estimate the direction of the forces along the fault that has caused the earthquake. This is called the study of earthquake-source mechanisms and was developed by J. Shida and H. Nakano in Japan and P. Byerly in the United States in the 1910s to 1920s, and in the 1930s by H. Honda in Japan.

If faulting occurs in the direction shown in Figure 3-7a the initial motion of the *P* wave (see Figure 1-4a) radiated by the faulting pushes the matter away from the source in the regions designated by a plus sign, and in those marked with a minus sign, the motion pulls the matter toward the source. If the direction of the faulting is reversed, as in Figure 3-7b, the radiation pattern is naturally reversed. By examining the mechanism of the earthquake source and then analyzing the direction of faulting, one should be able to deduce whether the fracture zone offsetting an oceanic ridge is a transform fault or not. In other words, one needs only to decide whether the actual faulting that takes place when earthquakes occur on the fracture zone is in the direction shown in Figure 3-5a or Figure 3-5b.

The progress that has been made in this field of study is due in great part to the data gathered by the World Wide Standard Seismograph Network (WWSSN), which the United States has set up throughout the world. After World War II, both the USSR and the

United States immensely improved their seismometer networks in order to be able to detect underground nuclear explosions. As a result, the field of seismology in both countries has made great progress. Actually, such seismological observation systems are contributing mostly to pure seismological studies rather than to the detection of nuclear explosions. The seismological data on oceanic ridges throughout the world, gathered for the first time by such a standardized observation system, provided L. Sykes of the Lamont-Doherty Geological Observatory with much of the material for his research in earthquake-source mechanisms.

It was already known that earthquakes in fracture zones occur only in the zone of displacement represented by bb' in Figure 3-5b—a fact that in itself strongly supports the concept of a transform fault. Sykes sought to clarify further the source mechanism of earthquakes in fracture zones. The result of his study was reported in 1966. It verified Wilson's prediction perfectly. For earthquakes that occurred along fracture zones between ridges, the direction of the faulting agreed with that predicted by Wilson without exception. The earthquakes that occur at the crests of oceanic ridges, according to Sykes, generate the waves one would expect from faulting caused by tensile forces as the two sides diverge from the ridge. It is hard to imagine a more direct or convincing proof of Wilson's theory of the transform fault.

## "Annual Rings" on the Ocean Floor:
## A Challenge to Geology

Supported by the Vine-Matthews-Morley hypothesis, which explained the striped geomagnetic anomaly pattern on the ocean floor, and by much additional evidence, the sea-floor spreading hypothesis became firmly established in the mid-1960s. It was at the Lamont-Doherty Geological Observatory that significant contributions to this development were made, for there, awaiting analysis and interpretation, were the results of decades of magnetic surveys covering all the oceans of the world. J. Heirtzler, W. C. Pitman, and others at the observatory undertook the arduous task of analyzing these data with computers. Soon, the geomagnetic anomaly profiles of all the earth's oceans had been systematically organized to provide a comprehensive description of the chronology of the sea floor.

The chronology of geomagnetic field reversals that can be deduced from paleomagnetic studies on land rocks covers only about four

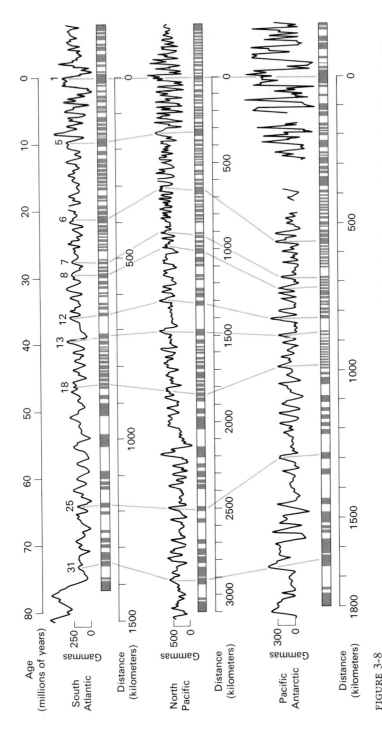

FIGURE 3-8

Geomagnetic anomaly profiles recorded over the floors of three oceans by ships traveling in a direction perpendicular to the mid-oceanic ridges. Note the similarity of the profiles, especially of the numbered peaks, which have easily recognizable shapes (each number refers to a particular lineation). The light and dark stripes along the horizontal bar beneath each profile indicate the succession of normally and reversely magnetized bodies of volcanic rock running parallel to the ridge. The spacing between magnetic bodies varies in each ocean because the spreading rates have been different (the rate in the South Atlantic is believed to have been the most constant), but each ocean has the same sequence of 171 reversals that extend back 76 million years, covering the entire Cenozoic period. [After J. R. Heirtzler et al., "Marine Magnetic Anomalies, Geomagnetic Field Reversals, and Motions of the Ocean Floor and Continents." *J. Geophys. Res.* **73**, p. 2119, 1968. Copyrighted by American Geophysical Union.]

million years, and contains only several epochs of geomagnetic reversals. However there are dozens of magnetic stripes across the sea floor. How, then, were we to infer the chronology of geomagnetic stripes prior to the past four million years? Heirtzler and his colleagues set out to estimate it by assuming that the rate of sea-floor spreading had been constant in the South Atlantic. As exemplified for three oceans in Figure 3-8, the geomagnetic profiles of the different major oceans showed a similar sequence of peaks and troughs for each. Many of the peaks and troughs were found to be common to all the oceans. Some of the peaks that were distinctive enough to be recognizable almost everywhere were then numbered to facilitate comparison of different profiles. Scientists at Lamont-Doherty were able to trace the magnetic anomaly profiles as far back as 76 million years ago (Figure 3-9). They found that in the course of this time 171 geomagnetic field reversals had occurred. Once this method of estimating the geomagnetic time scale proved workable, the age of the ocean floor, wherever the floor displayed the striped geomagnetic anomaly pattern, could be traced as far back as one wished. The striped pattern, in other words, provides "annual rings" for the ocean floor. (Of course, on the ocean floor, each stripe represents growth that has taken place, not in the course of a single year but during periods of time varying from 20,000 to several million years.) Heirtzler's assumption that the rate of sea-floor spreading in the South Atlantic had been constant was subsequently supported by the evidence gathered by the Deep Sea Drilling Project (DSDP), to be described in the next section of this chapter.

By measuring striped patterns across the entirety of the vast spread of the ocean floor, it is possible in principle to construct magnetic lineations, or *isochrons*—which connect those points of the crust having the same age. The result is a geological map of the ocean floor. A geological map of the land may be more detailed in its description of such items as the nature of the rocks, but the mapping depends largely on the hammers, other manual equipment, and sheer physical endurance of large numbers of geologists who are dedicated to the step-by-step examination of the earth. By contrast, the geological mapping of the ocean floor can be done rapidly by research vessels with magnetometers trailing behind them. The results of such surveys are shown in the up-to-date compilation of marine magnetic findings plotted in Figure 3-10. The shaded areas represent those portions that Heirtzler and his colleagues succeeded in plotting as early as 1968. Explanations of lineations outside the shaded areas will be given in Chapter 4.

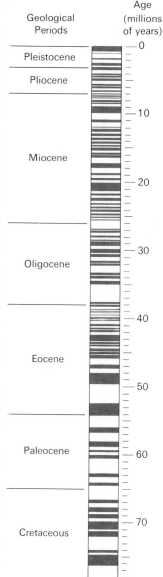

| Geological Periods | Age (millions of years) |
|---|---|
| Pleistocene | 0 |
| Pliocene | |
| Miocene | 10 |
| | 20 |
| Oligocene | 30 |
| Eocene | 40 |
| | 50 |
| Paleocene | 60 |
| Cretaceous | 70 |

FIGURE 3-9
The chronology of geomagnetic field reversals for the past 76 million years. [After J. R. Heirtzler et al., "Marine Magnetic Anomalies, Geomagnetic Field Reversals, and Motions of the Ocean Floor and Continents." *J. Geophys. Res.* **73**, p. 2119, 1968.]

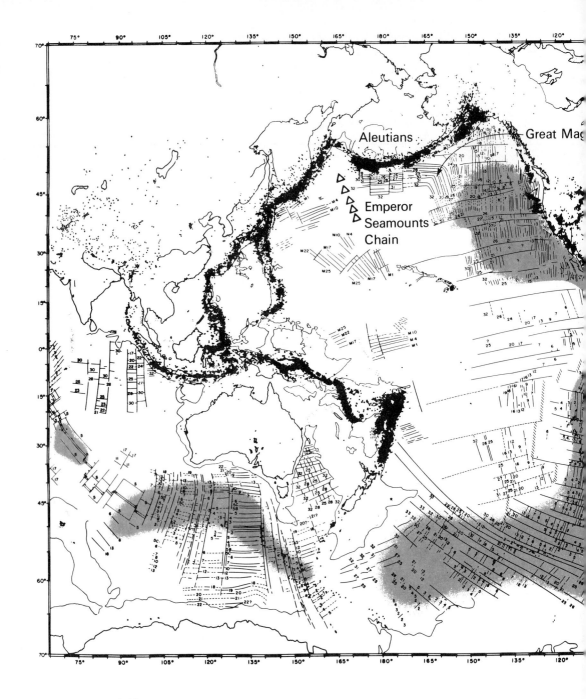

FIGURE 3-10
Isochron map of the ocean floor. The numbers along each isochron give the anomaly numbers (see text). Shaded sections represent the areas for which data were compiled in 1968 by J. Heirtzler and

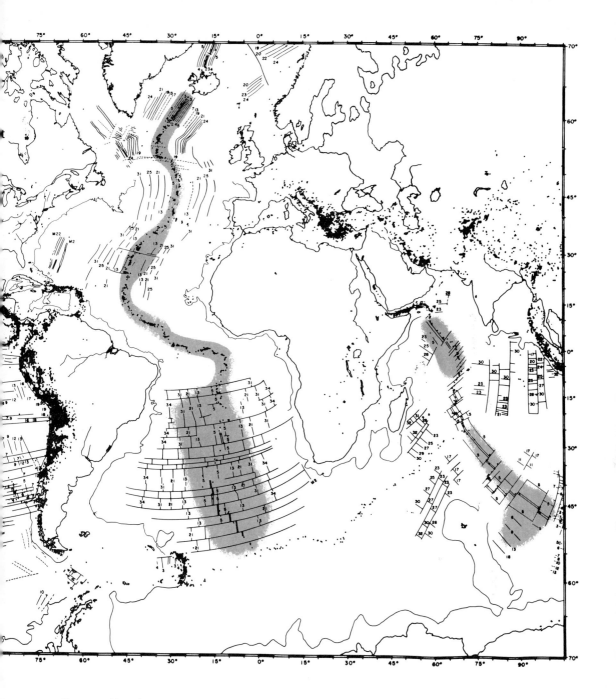

colleagues. Also shown are the mid-oceanic ridges, transform faults, and the distribution of epicenters throughout the world. [Compiled by W. C. Pitman, III, R. L. Larson, and E. M. Herron, 1974. Reproduced with permission of the authors and the Geological Society of America.]

Although isochrons have been plotted for large portions of the ocean floor, other regions are blank, such as much of the central and western Pacific and regions in the northern Atlantic far from the oceanic ridges. The isochrons have not been plotted, either because detailed magnetic studies had not yet been made, or because magnetic surveys revealed no striped pattern of geomagnetic anomalies. Those areas revealing no magnetic patterns are called *magnetic quiet zones*, and the question of what caused them has several possible answers. It may be that the speed of sea-floor spreading was extremely rapid at the time, thus producing a larger area of sea floor while the geomagnetic field was facing in one direction. Or perhaps no geomagnetic field reversal occurred for quite a long period of time; paleomagnetic studies on land reveal that such periods, devoid of magnetic reversals, existed at the end of the Paleozoic era and in the Jurassic and Cretaceous periods. Other possible causes for quiet zones are that the ocean floor may have failed to record the geomagnetic reversals or lost its remanent magnetism because of secondary factors, or that the ocean floor of a magnetic quiet zone may have been produced by a process totally different from the sea-floor spreading process that is currently assumed. This problem was an important one throughout the late 1960s, and it was only after several years that a solution was attained, as will be explained in Chapter 4, where the riddle of magnetic lineations in the unshaded areas in Figure 3-10 is attacked.

**The Deep Sea Drilling Project**

Even the apparent success of the sea-floor spreading hypothesis did not stop man's desire for further endeavor. Since they were not entirely satisfied with the magnetic method of determining the age of the ocean floor (which, though logical, is based on a chain of hypotheses), scientists sought to develop a method of more direct sampling for this purpose. The Deep Sea Drilling Project (DSDP), initiated in the United States in 1968, fulfilled this goal. The objective was to gain direct information on the structure and history of the ocean floor by drilling to the basement rock and collecting samples of all the sediments covering the ocean bottom. Since this was to be a huge undertaking, too big for any single institution, a consortium of leading oceanographic institutions was formed. The consortium, consisting of Scripps Institution of Oceanography, Lamont-Doherty Geological Observatory, the University of Washington, the Univer-

FIGURE 3-11
The Deep Sea Drilling Project vessel *Glomar Challenger* (10,500 tons). [Courtesy Deep Sea Drilling Project.]

sity of Miami, and the Woods Hole Oceanographic Institution, was called JOIDES (Joint Oceanographic Institutions Deep Earth Sampling). In order to accomplish the difficult task of boring holes in the sea floor, which lies several thousand meters below the surface of the ocean, scientists refined the boring technique used for offshore oil-well drilling and built a new drilling ship, the *Glomar Challenger* (Figure 3-11). The ship has conducted her drilling operations since August 1968, covering the Atlantic, Pacific, Indian, and Antarctic Oceans. Figure 3-12 shows the sites at which drilling was completed by October 1973. Each drilling cruise consists of a two-month "leg." When the DSDP terminated in 1975, 44 legs had been completed.

The most remarkable achievement in the early phase of DSDP was the set of results obtained in the Atlantic on the *Glomar Challenger's* Leg III. Holes were drilled through to the basaltic rocks thought to form the original volcanic sea floor. The sediments just above the basalts were believed to be only slightly younger than the

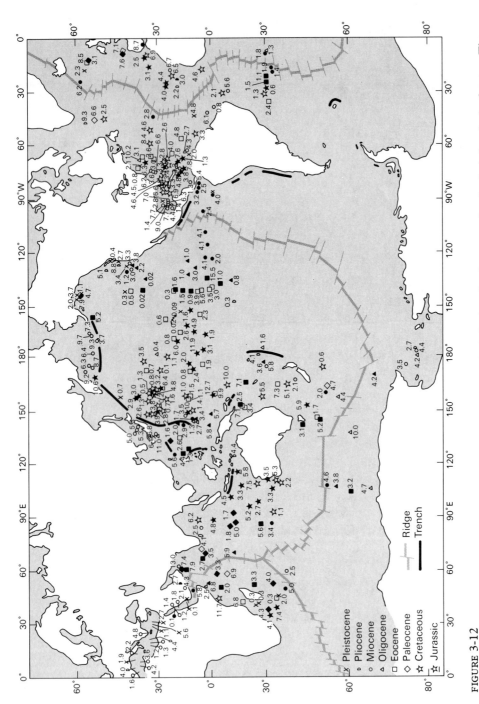

FIGURE 3-12
Deep Sea Drilling Project holes, legs I through XXXII. Numbers at the hole sites give the penetration depths in hundreds of meters. The symbols designate the age of the sediment by epoch or period. Black symbols at a site indicate that basalt was reached. The contour is for a depth of 4000 meters. [After K. Koizumi and S. Uyeda, "Earth Science and Deep Sea Drilling Project." Reproduced from *Kagaku* **44**, 4, p. 203, 1974. Copyright © 1974 by K. Koizumi. Courtesy of Iwanami Shoten Publishers, Tokyo.]

Ridge
Trench

× Pleistocene
◻ Pliocene
○ Miocene
△ Oligocene
◻ Eocene
● Paleocene
◇ Cretaceous
☆ Jurassic

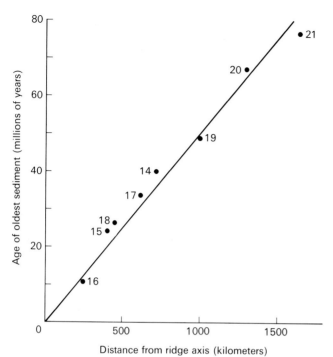

FIGURE 3-13
The results from the DSDP, Leg III, for sites 14 through 21
across the Mid-Atlantic Ridge at approximately 30 degrees south.
The age of the oldest sediment at each site is plotted against the
distance of the site from the ridge axis, revealing a strongly linear
relationship between the two factors. This relationship suggests a
constant spreading rate of 2 centimeters per year throughout the
Cenozoic period. [After A. E. Maxwell and R. von Herzen, "The
*Glomar Challenger* Completes Atlantic Track—Highlights of Leg
III." *Ocean Industry* **4**, 5, p. 64, 1969.]

basalt, and the ages of these sediments were determined from the
fossils in them. These ages were found to be in excellent agreement
with those one would infer from the distance of the site from the
Mid-Atlantic Ridge (in accord with the sea-floor spreading hypothesis), as shown in Figure 3-13. The age of the sea floor of the
Pacific Ocean, as indicated by the magnetic stripes, also coincided
with that determined by the Deep Sea Drilling Project. Where the
ages of magnetic lineations were unknown, holes drilled in critical areas provided the key information. It was confirmed that
the youngest part of the Pacific Ocean floor is in the eastern Pacific
and that the age generally increases toward the west, as shown in
Figure 3-14.

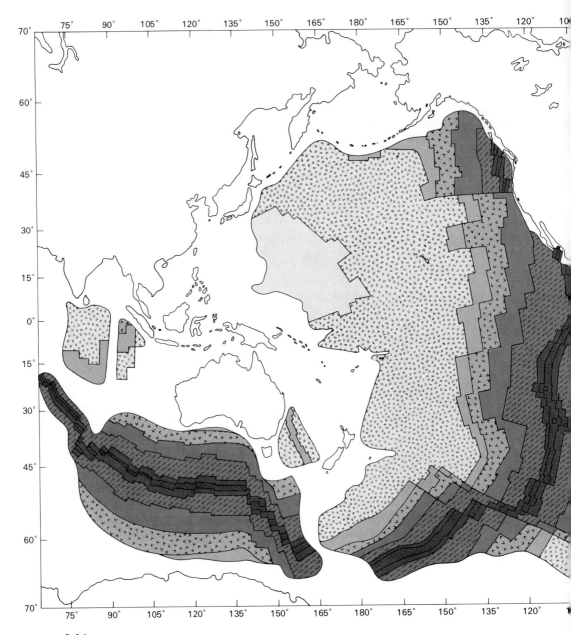

FIGURE 3-14
The age of the ocean basin. [Compiled by W. C. Pitman, III, R. L. Larson, and E. M. Herron, 1974.
Reproduced with permission of the authors and the Geological Society of America.]

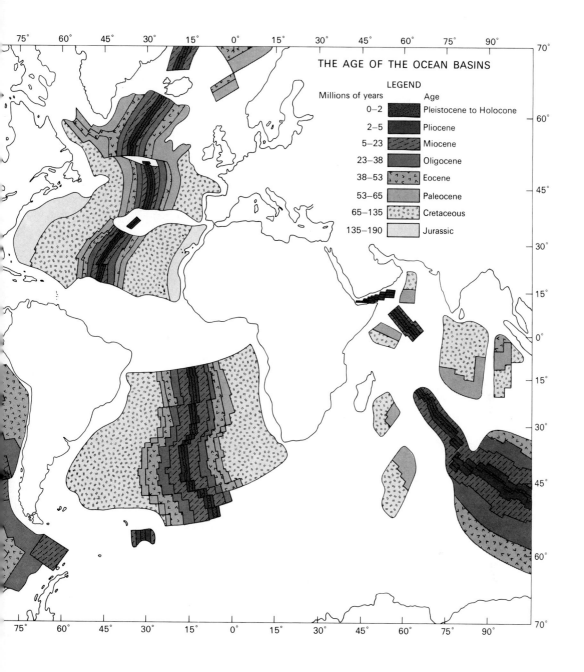

THE AGE OF THE OCEAN BASINS

LEGEND

| Millions of years | | Age |
|---|---|---|
| 0–2 | | Pleistocene to Holocone |
| 2–5 | | Pliocene |
| 5–23 | | Miocene |
| 23–38 | | Oligocene |
| 38–53 | | Eocene |
| 53–65 | | Paleocene |
| 65–135 | | Cretaceous |
| 135–190 | | Jurassic |

Although we will not go into more details here, the Deep Sea Drilling Project has provided an enormous amount of valuable information about the history of the oceans. Especially important from a geophysical perspective was verification that the second layer of the oceanic crust immediately beneath the layer of sediments is basalt, at least at the top. What, then, is the composition of the deeper part of the second layer, and of the third layer? Although the DSDP was terminated in 1975, scientific interest remained so keen that the project has been maintained as an international one called IPOD (the International Phase of Ocean Drilling). Thus very active drilling operations are still being conducted in an effort to understand the origin and evolution of the ocean bottom. JOIDES is now joined by institutions from the USSR, the Federal Republic of Germany, France, Japan, and the United Kingdom, as well as by several more American institutions. Man's desire for knowledge is boundless.

## The Mysteries of the Pacific Ocean Floor

Figure 3-15 is a part of the map of magnetic anomalies around Japan that a group of us prepared in 1966. A distinct striped pattern is discernible in the areas east of the island arc of Japan—stripes that extend from northeast to southwest. At first we considered this striped pattern of magnetic anomalies to be the oldest one in the Pacific: born at the East Pacific Rise, it had moved across the vast Pacific Ocean and was about to descend beneath the Kurile Trench. This would mean the stripes should be progressively older to the north.

Such a simple assumption was challenged, however, as American scientists extended the survey of magnetic anomalies in the East Pacific northward to the area off the Canadian coast, the Gulf of Alaska, and the Aleutians. The magnetic stripes in the East Pacific, which extend from north to south in the southern areas, turn almost at right angles off the Alaskan coast to run from east to west (see Figure 3-10). In the northeast Pacific the magnetic stripes get progressively younger as they extend eastward, demonstrating that in this area the ridge that produced the ocean floor once existed to the east of those stripes. If we follow the same stripes northward around their bend to the point at which they run east-west off the Alaskan coast and the Aleutian Islands, it would seem logical that they would get older to the south and younger to the north.

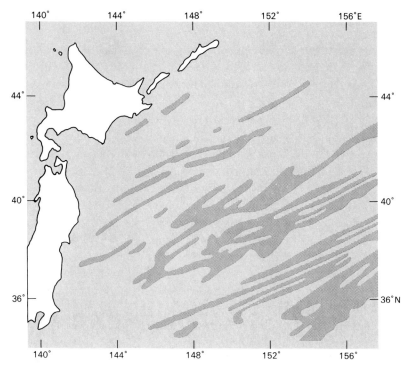

FIGURE 3-15
Total magnetic intensity anomalies in the northwestern Pacific. The shaded areas
are positive in anomaly, and the white areas, negative. [After S. Uyeda et al.,
"Results of Geomagnetic Survey During the Cruise of R/V *Argo* in Western
Pacific 1966 and the Compilation of Magnetic Charts of the Same Area." *Bull.
Earthquake Res. Inst.* **45,** p. 799, (1967).]

However, such a deduction seemed difficult to accept because to
the north lies the Aleutian Trench, into which the sea floor is sup-
posed to be descending. In contrast, the sea-floor spreading hypothesis would seem to require that the closer the ocean floor is to a
trench, the *older* it should become. Therefore, shouldn't the stripes
get younger toward the south and older toward the north, where the
Aleutian Trench is? The east-west magnetic stripes south of the
Aleutians extend westward until they almost join the magnetic
stripes of the western Pacific off Japan. If the Japanese and Aleutian
lineations are continuous, the problem mentioned above for the
Aleutian area also applies to the Japanese area. Despite this difficulty, it seemed also unlikely that the magnetic anomalies, which
form such a continuous striped pattern across all of the other ocean
floors, should end so abruptly in the area between the Aleutians and

Japan (see Figure 3-10). More than one mystery, then, concerning the magnetic stripes in the northern Pacific remained to be solved.

The north Pacific was not the only area that posed unanswered questions, however; the age and area of the west Pacific were also still unknown by 1968, as indicated by the vast unshaded area in Figure 3-10. The west Pacific is known for its abundant flat-topped submarine mountains or seamounts (called *guyots*) and coral reefs (see Figure 2-2). For this reason the physiography of the sea floor of the western Pacific appears to be very different from that of the eastern Pacific. Did the two parts of the Pacific ocean floor originate from the East Pacific Rise by the same process of sea-floor spreading? Some scientists maintained that an extensive undersea rise had once existed in the area of the western seamounts. For example, the late H. Hess and H. W. Menard called this hypothetical rise the Darwin Rise. However the former existence of this rise would be evidenced by a large-scale depression in the same area, and the presence of such a depression has recently been subjected to serious doubt by the Deep Sea Drilling Project.

The sea-floor spreading hypothesis, then, was supported by observational data from the spreading-center areas—the ridges—in the 1960s. But sea floor away from the ridges and close to the trenches, such as the western Pacific, still posed a number of riddles. After introducing the concept of plate tectonics, we will come back to these questions in the following chapters.

# Chapter 4

# Plate Tectonics

プレート・テクトニクス

## Plate: A New Concept

Once the sea-floor spreading hypothesis began to prove itself viable, almost everyone was attracted to it. Geophysicists began to greet one another with the question, "Do you believe in sea-floor spreading?" And they found themselves answering, "Yes." In 1967–68, special sessions on the sea-floor spreading hypothesis were held at geophysical meetings throughout the world, and hundreds of papers and reports were submitted, most of them attempting to demonstrate how effectively the sea-floor spreading hypothesis explained the phenomenon that each scientist was studying. Among the countries actively involved in solid earth science, only Russia and Japan appeared to remain skeptical. In most other countries the hypothesis was extremely popular. Once the transform fault hypothesis was enthusiastically accepted, many scientists were eager to explore the ultimate logic of the sea-floor spreading hypothesis.

Now let us examine once again the distribution of earthquake epicenters shown in Figures 2-4 and 3-10. It seems obvious that the earthquakes occur mostly along oceanic ridges, along transform faults and island arcs, and in orogenic belts like the Andes and the Alpine-Himalayan region. In contrast only a few earthquake foci are scattered over the vast areas that are surrounded by these earthquake belts. Since earthquakes are thought to originate on faults that are ruptured as a result of accumulated stress, the areas without earthquake foci are either devoid of such stress or are incapable of forming faults even under stress. Although it is known that rocks tend to lose their brittleness and become ductile under high temperature and pressure, it seems unlikely that the concentration of earthquakes in narrow belts is because these belts are cold and brittle whereas the rest of the crust is warm and ductile. A more plausible explanation is that the belts of epicenters divide the earth's surface into blocks that are strong and fairly rigid, and that these blocks move relative to each other, causing earthquakes along their margins while leaving the inner structure of the blocks intact.

It had been suggested that the earth's surface layer is highly rigid to a depth of about 70 kilometers, and that a softer layer lies below. This conjecture was the result of highly sophisticated studies of the propagation of seismic waves made during the 1960s. F. Press synthesized all of the available information to arrive at his discovery of the remarkable variation in the velocity of seismic shear waves that prevails at different depths, shown in Figure 4-1. To generate his model of the earth, of which only the uppermost part is shown in Figure 4-1, Press used what is called a Monte Carlo method. In this method a high-speed electronic computer seeks, in a series of random trials, models that will satisfy various geophysical observations. In this model, the $S$-wave velocity jumps from about 3.6 kilometers per second to 4.6 kilometers per second at a very shallow depth that corresponds to the Moho discontinuity. The velocity keeps increasing with depth to about 70 kilometers where it drops to about 4.2 kilometers per second. Deeper in the mantle the velocity gradually increases again. Wave velocity should increase with pressure under normal circumstances: therefore, something unusual must be occurring at the depth of 70 kilometers. Don Anderson and others proposed a convincing explanation for the cause of this decrease. They attributed it to the partial melting of the mantle.

Mantle material is a composite of silicates with extremely complex melting properties—so complex that petrologists, after an enormous amount of research, are just beginning to understand them. We do know, however, that the mantle does not melt completely at a single temperature, as ice does. Instead the melting takes place within a certain temperature range. In the beginning of the melting process only a small part of the material melts. This phenomenon of partial melting causes a drop in seismic-wave velocity, especially in the shear velocity. The temperature at which melting begins is called the *solidus temperature*. It can be easily inferred from the seismic data that partial melting has softened the mantle at a depth between 70 and 260 kilometers. The softened layer, the *asthenosphere*, lies at a

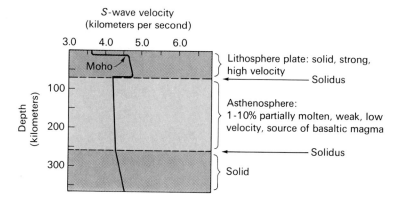

FIGURE 4-1
A modern view of the structure of the outermost layers of the earth is illustrated by
a plot of S-wave velocity against depth. Note how velocity changes at the
70-kilometer depth mark the boundary between the lithosphere and the partially
molten asthenosphere. [After F. Press and R. Siever, *Earth.* W. H. Freeman and
Company. Copyright © 1974.]

depth of approximately 70 kilometers beneath the rigid outer layer,
the *lithosphere.*

Once this information has been assembled, the earth's outer layers
can be envisioned as a rigid lithosphere, consisting of several blocks
or plates, which covers an underlying softer asthenosphere. Both the
drifting of the continents and the spreading of the sea-floor may be
ascribed to the movements of these rigid plates. Furthermore, their
interaction is believed to be the cause not only of earthquakes, but
also of many other important phenomena on the earth's surface, such
as volcanic activity, oceanic trenches, and oceanic ridges.

This hypothesis, which has come to be known as the theory of
*plate tectonics,* was advanced by D. McKenzie and R. Parker (1967)
and independently by W. J. Morgan (1968). X. Le Pichon, a French
oceanographer working at Lamont-Doherty Geologic Observatory,
was one of the first to see the importance of this theory and quickly
applied it to reveal the movements between most of the larger plates.
We will now examine the remarkably simple but equally profound
ideas advanced by these young scientists.

## Testing the Hypothesis

There are three major types of plate boundaries, as Figure 4-2 clearly
shows. One is the boundary created when two plates are moving apart
from one another. Typically, it is found at a mid-oceanic ridge where

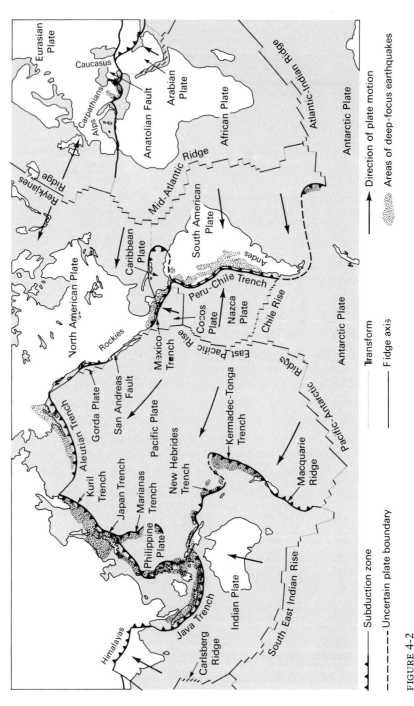

FIGURE 4-2

The earth's lithosphere is broken into large rigid plates, each moving as a distinct unit. The relative motions of the plates, assuming the African plate to be stationary, are shown by the arrows. Plate boundaries are outlined by earthquake belts. Plates separate along the axes of mid-ocean ridges, slide past each other along transform faults, and collide at subduction zones. [After J. F. Dewey, "Plate Tectonics." Copyright © 1972 by Scientific American, Inc. All rights reserved] new plates are being formed. It is called an *accreting,* or *diverging boundary.* Another type of boundary is created where two plates are moving toward one another. This is called a *converging boundary.* Oceanic trenches and certain mountain ranges, such as the Himalayas, are located along such boundaries. At the trenches, oceanic plates are thought to descend, or *subduct,* into the earth, whereas at the Himalayas two continental plates are in collision. Boundaries that occur along trenches are also called *consuming boundaries.* The third type of plate boundary occurs along transform faults, where the relative plate motion is parallel to the boundary.

Let us now consider two rigid plates A and B on a globe, as shown in Figure 4-3a, which are separated by oceanic ridges and transform faults at one boundary and by oceanic trenches and transform faults at the other. Geometrically, the only possible movement of the rigid plates is in a direction parallel to the transform faults, and such movement can be described as a rotation around a pole, $_AP_B$. Transform faults, which indicate the directions of such relative motion between plates (A and B in this example), lie on latitude circles around the pole $_AP_B$. The ridges, which are usually at right angles to the transforms, thus lie along the longitudinal, or meridional, circles. Since most trenches do not strike perpendicular to the transform faults, they do not lie meridionally. Now, if we draw a map on a Mercator projection, with $_AP_B$ as the pole, all the transform faults between plates A and B will lie parallel to the lines of latitude and the ridges will lie parallel to the lines of longitude as in Figure 4-3b. The direction of the relative motion between two plates can also be assessed from the first motion of an earthquake occurring at the plate boundary. The direction of seismic slip should also lie parallel to the latitude circle (for the same reason the transform fault does), as shown by the arrows in Figure 4-3b.

D. McKenzie and R. Parker (1967) studied earthquake-source mechanisms along the boundary of the Pacific Plate, with this idea in mind: if indeed the interactions of rigid plates cause earthquakes, an analysis of earthquakes along trenches, transforms, and ridges bordering the Pacific Plate should all reveal the same direction of motion for the Pacific Plate in relation to the neighboring plate. They found that this was true, and that the location of the pole describing the motion of the Pacific Plate relative to the North American Plate was at 50° north, 85° west (Figure 4-4). They obtained the position of this pole of rotation from the direction of the San Andreas Fault—a transform fault—and the average direction of fault motion of the aftershocks in the Kodiak Island region that followed the Great

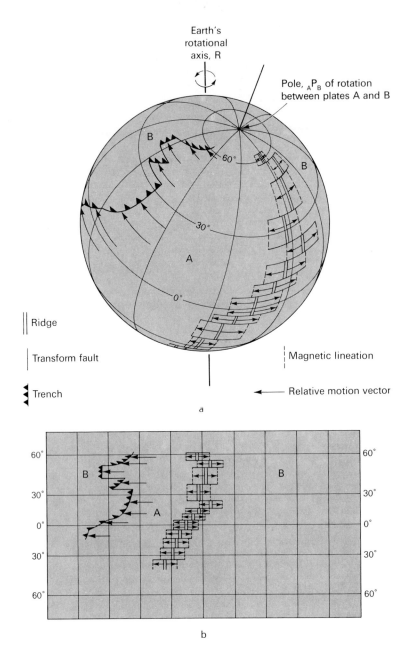

FIGURE 4-3
(a) The rotation of rigid plates on a sphere. At right is a spreading ridge with transform faults. At left is a plate boundary formed by trenches and transform faults. $_AP_B$ is the pole of relative rotation between plates A and B, which should not be confused with R, the pole of the earth's rotation. The arrows indicate both direction and speed of relative motion. Note that the spreading rate, indicated by the length of each arrow, is larger at lower latitudes, and the direction of the relative motion is parallel to the transform faults.
(b) A Mercator projection of the globe shown in (a). Projection is made on the pole $_AP_B$. Note that all transform faults and relative-motion vectors are now parallel to latitude lines, and that ridges and magnetic lineations are parallel to longitude lines. The width of magnetic lineation and the length of the arrows are now independent of latitude.

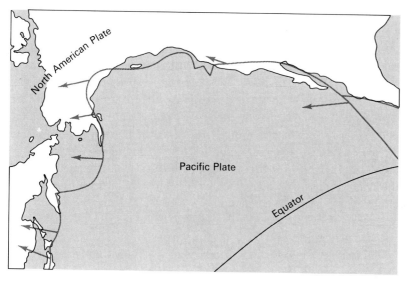

FIGURE 4-4
A Mercator projection of the Pacific with a pole at 50° North, 85° West. The arrows show the direction of motion of the Pacific Plate relative to the North American Plate. If both plates are rigid, all relative-motion vectors must be parallel to each other and to the upper and lower boundaries of the figure that are parallel to the circles of latitude. [After D. P. McKenzie and R. L. Parker, "The North Pacific: An Example of Tectonics on a Sphere." *Nature* **216,** p. 1276, 1967.]

Alaskan Earthquake of 1964. (The pole should be located at the intersection of great circles taken perpendicular to these two directions, if the plate hypothesis is correct.)

In the meantime, W. J. Morgan was thoroughly examining the observed widths of the striped patterns of geomagnetic anomaly and the direction of transform faults. Let us examine Figure 4-3a once again. Now, if the rigid-plate hypothesis is correct, one can see that (given the geometric requirement that the only possible direction of plate movement is parallel to the transform faults) the speed of the relative movement between two plates, or, in other words, the speed of both sea-floor spreading and subduction, represented by the lengths of the arrows in Figure 4-3a, should increase with increasing distance from the pole $_A P_B$. It can be shown, in fact, that the speed should be proportional to the cosine of the angle of latitude. Hence, the relative plate movement at the pole $_A P_B$ will be zero, while the spreading rate will be maximum at a distance 90° from $_A P_B$. According to the sea-floor spreading and Vine-Matthews-Morley hypotheses, the spreading rate can be determined by the width of the magnetic stripes. So, theoretically, the pole of rotation between two plates can be estimated also from the distribution of the width of magnetic stripes. In other words, the pole can be estimated independently from both the distribution of the directions of transform faults, or seismic first-motion data, and from the distribution of the widths of magnetic stripes. If the pole of rotation is correctly estimated, the magnetic stripes should look like the dashed lines in Figure 4-3a. They are parallel to the meridional lines and they decrease in width toward the pole $_AP_B$ as the spreading speed (represented by the lengths of the arrows) decreases toward the pole. Here, it may be seen that all of these magnetic stripes will become parallel and of constant width in the Mercator projection in Figure 4-3b, just as all the longitudinal circles that converge at $_AP_B$ in Figure 4-3a are converted to the north-south parallel lines in Figure 4-3b. The lengths of arrows also become constant.

These relations, which are intrinsic to the Mercator projection, can be used to check the validity of plate tectonic concepts graphically. Figure 4-5 is such an example. It shows a Mercator projection of the world map drawn about a pole of rotation. The position of the pole was determined from the distribution of the sea-floor spreading rates—that is, the width of magnetic stripes—in the East Pacific Rise. In this map all the transform faults of the East Pacific Rise do indeed lie parallel to the lines of latitude. This indicates that the pole that has been determined from the distribution of the directions of transform faults coincides with the pole that has been determined from the distribution of the spreading rate. Thus it now seems almost certain that the Pacific Plate and the plate east of the East Pacific Rise (the Antarctic Plate) have undergone relative rotation about this pole since at least the time at which the anomaly stripe used for the analysis was formed (70 million years ago). These findings were widely read by earth scientists before their publication in about 1967, and were the source of lively debate. Now they started to greet one another with the query, "Do you believe in plate tectonics?"

It is intriguing to note, in Figure 4-5, that the direction of many of the faults to the north, which H. W. Menard calls "fossil faults," do not lie parallel to the lines of latitude but are off by 30°. Would this mean that, long ago, the relative motion of rotation in the northeast Pacific, as represented by these faults, was not the same as that of the plates in the south Pacific? This was perplexing to some. However, the discrepancy is not so puzzling when one realizes that the motions deduced from the present methods are strictly the *relative* motions between two neighboring plates. In discussing motion along the San

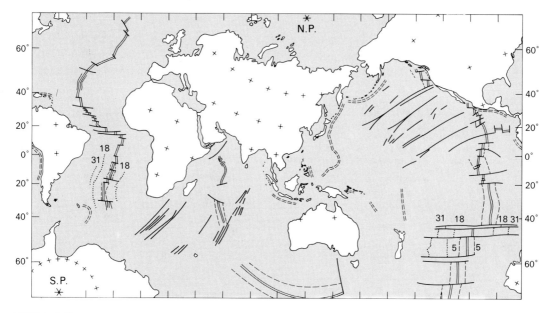

FIGURE 4-5
A Mercator map of the world drawn about the pole of rotation of the plates in the Southeast Pacific. The numbers at the magnetic stripes are anomaly numbers. Anomalies 5, 18, and 31 correspond to the ages 10, 45, and 70 millions of years ago. [After X. Le Pichon, "Sea-Floor Spreading and Continental Drift," *J. Geophys. Res.* **73,** p. 3661, 1968. Copyrighted by American Geophysical Union.]

Andreas Fault, for example, it is correct to say either, "the Pacific Plate is moving northwest relative to the North American Plate," or "the North American Plate is moving southeast relative to the Pacific Plate." The direction of the fault reveals the direction the plates are moving relative to each other, but it doesn't tell us that one plate is moving and the other is stationary. Therefore a plate can remain rigid and also have nonparallel sets of transform faults when portions of the boundary face different plates (Figure 4-6). In earlier times the northern and southern parts of the Pacific Plate *did* face different plates, and the transforms along the two boundaries had different directions. The plate that was bounded by the northern part of the East Pacific Rise was not the same plate as the one in the south, and it was consumed and disappeared some 30 million years ago when that part of the East Pacific Rise collided with the North American Plate. This remarkable story will be told in the following sections of this chapter.

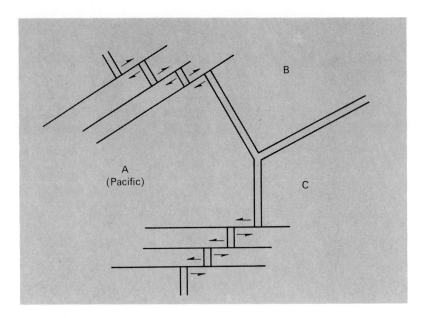

FIGURE 4-6
Schematic diagram showing nonparallel transform faults can be formed when one plate (A) has more than one neighboring plate (B and C).

## The Migration of Oceanic Ridges
## in the North Pacific

Recall that the magnetic stripes in the east Pacific bend abruptly in the Gulf of Alaska (Figure 3-10). This bend was discovered by the American scientist G. Peter and his colleagues during 1966 and 1967, and was named the *Great Magnetic Bight*. They extended their survey area westward and traced the lineations trending in an eastwest direction up to the Emperor Seamounts at the western end of the Aleutian Islands. Since the east-west lineations south of the Aleutians are the continuation of those in the southeast Pacific, they presumably become younger as they approach the Aleutian Trench. As we stressed toward the end of the preceding chapter, this pattern appeared to violate the basic concept of sea-floor spreading—the postulate that new sea floor is formed at the ridges and spreads toward the trenches.

Part of this dilemma had already been resolved in 1966 by Vine in his analysis of stripes off the coasts of California, Oregon, and Washington. There the stripes get younger eastward, again as they approach the location of an ancient trench that is no longer active. Vine reasoned that a rise ancestral to the East Pacific Rise once lay off the coast of California, and was separated from the coast by a trench. The rise migrated toward the trench until eventually the two met and annihilated each other. Lithosphere that had formed east of the rise has been subducted beneath North America. On lithosphere formed west of the ancient rise, the magnetic stripes get younger to the east. The reason for this, Vine explained, was that originally the stripes were getting younger toward the rise, in accord with the theory of sea-floor spreading; however, the rise has now vanished in the ancient trench.

Following Vine's lead, in 1968, W. C. Pitman and D. Hayes proposed a strange yet plausible explanation for the enigma of the stripes south of the Aleutians (Figure 4-7). They maintained that in the late Cretaceous period (around 75 million years ago) there were three active oceanic ridges in the northeast Pacific radiating from a common point (Figure 4-7a). The sea floor was spreading away from the crest of each ridge. (In the figure the trenches are presented merely for the purpose of reference—their exact position is not known.) In order to examine the process that took place when the three plates were being produced by the three oceanic ridges, let us suppose that plate I, representing the Pacific Basin, is fixed. This is entirely legitimate, because this way all the motions are taken to be relative to plate I. Since each plate was growing along its ridge boundaries, with plate I held immobile, it is necessary that plates II and IV must have migrated away from plate I toward the north and east respectively. At the same time the ridge separating plates I and II was migrating northward and the ridge separating plates I and IV was migrating eastward, adding new ocean floor to plate I (Figure 4-7b). Finally the ridge between plates I and II disappeared into the Aleutian Trench (Figure 4-7c), while the ridge between plates I and IV descended beneath the American continent, leaving behind magnetic stripes that are younger toward the north and east (Figure 4-7d), in accord with Vine's interpretation to the south. Ingenious as this explanation is from a geometrical point of view, the idea that oceanic ridges should descend into oceanic trenches seemed a bit strained. Does the rising rim of a convection cell sink into the descending rim? This point was vehemently criticized by the opponents of the sea-floor spreading hypothesis. We will come back to this problem later.

The possibility that the oceanic ridge itself migrates is quite intriguing. In fact, this suggestion had been made earlier, prompted by certain observations. The African continent, as Figure 4-2 shows, is

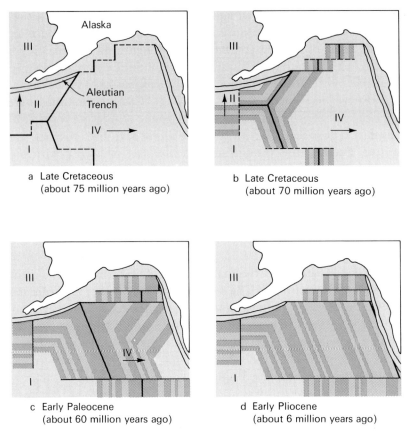

a  Late Cretaceous
(about 75 million years ago)

b  Late Cretaceous
(about 70 million years ago)

c  Early Paleocene
(about 60 million years ago)

d  Early Pliocene
(about 6 million years ago)

FIGURE 4-7
Schematic diagrams of four stages in the development of the lineations of the Northeast Pacific. The arrows indicate the motions relative to Plate I (the Pacific Plate). [After W. C. Pitman and D. E. Hayes, "Sea-Floor Spreading in the Gulf of Alaska." *J. Geophys. Res.* **73**, p. 6571, 1968. Copyrighted by American Geophysical Union.]

surrounded by oceanic ridges in three directions. The Antarctic continent is even more completely surrounded by spreading ridges. Given these conditions, it would be geometrically impossible for new sea floor to be produced at all these ridges and to spread toward the continents unless, of course, there are trenches along the coasts that consume the sea floor as it spreads. There are no trenches around these continents, however. The oceanic ridges surrounding the African and Antarctic continents, then, *must* be migrating farther and farther away from those continents as they produce more and more sea floor.

An interesting corollary of plate tectonics is the geometrical requirement that there must be points at which three plates meet. These points have been named *triple junctions* by D. McKenzie and W. J. Morgan (1969). Depending on the types of boundaries meeting at the point of intersection there can be various types of triple junctions. The point at which three ridges meet in the Gulf of Alaska, in Figure 4-7, is an example. This type of junction is called an R-R-R junction, R symbolizing ridge. Similarly, all the combinations of R, T (trench), and F (transform fault) can form triple junctions, such as T-T-T, F-F-F, R-T-F, and so on.

In their important paper in 1969 on the theory of triple junctions, McKenzie and Morgan demonstrated that, on the basis of simple geometrical considerations, some types of triple junctions are stable whereas others are not: when the same plate configuration can be maintained through time, the triple junction is stable. In addition, they determined the conditions under which the velocities of each plate combine to make stable triple junctions. They pointed out that, even though the motions of each plate remain unchanged, the evolution of the triple junction can alter the nature (or type) of the plate boundaries. Such an alteration, they suggested, may be an important cause of tectonic changes over time. An important and dramatic example of this assertion was provided in the tectonic history of western North America, as will be shown in the next section.

## An Oceanic Ridge That Collided with the North American Continent

Let us examine Figure 4-8. The most active earthquake and volcanic zones in the world are in an area known as the *circum-Pacific orogenic belt*. These zones include the Aleutian Islands in the north, the Kuriles, the islands of Japan, the Izu-Bonin and Marianas Islands, down through the Ryukyus, the Philippines, and New Guinea, and the Tonga and Kermadec Islands and New Zealand in the south; they include also the west coasts of Central and South America. Although the prefix "circum-" implies an unbroken circle around the Pacific Ocean, there are breaks in the orogenic belt along the west coast of the North American continent and between Australia and the Antarctica. These two regions are also distinguished in Figure 4-8 as areas devoid of deep- and intermediate-focus earthquakes.

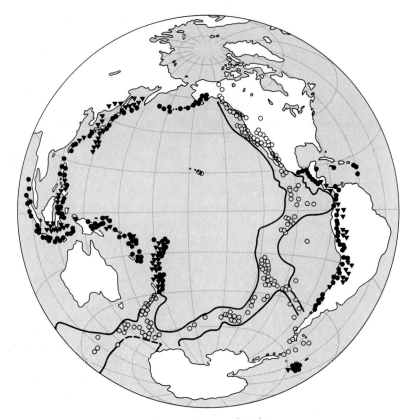

o Shallow-focus earthquakes
● Intermediate-focus earthquakes
▼ Deep-focus earthquakes

FIGURE 4-8
The circum-Pacific belt is interrupted by the East Pacific Rise. [After R. W.
Girdler, "Research Note—How Genuine Is the Circum-Pacific Belt?" *Geophys. J.*
**8,** p. 537, 1964.]

    R. Girdler of England pointed out in 1964 that "circum-Pacific" is
not an appropriate term because the circle is broken. He explained
this observation by stating that, whereas the ocean floor is descending
on both sides of the Pacific, a mid-oceanic ridge exists on the west
coast of the North American continent and in the area between Australia and the Antarctica. This fact was already known but it received
renewed attention, thanks to Girdler's insistence. Recall that on the
west coast of the North American continent there is no trench, nor
are there deep-focus earthquakes. The only notable feature is the San

Andreas Fault, along which shallow earthquakes occur. As a matter of fact, the San Andreas Fault is a transform fault that links the East Pacific Rise with the Juan de Fuca and Gorda Ridges (see Figure 3-6). It is a gap in the presently active orogenic belt. In the same region, however, there are in the coastal ranges extensive belts of sedimentary deposits that show that a trench once existed there like the one existing along the west coast of South America today.

What we must note again here is that the oceanic ridges themselves have migrated, so that the magnetic lineations off North America represent only one side of the East Pacific Rise. We recall Vine's explanation that the East Pacific Rise was once situated in the northeastern Pacific, producing ocean floor as it is doing in the southeastern Pacific today. The spreading ocean floor advanced toward the east and descended beneath a then existing trench off the coast of North America. This explanation certainly appeared to be helpful to an understanding of the geological structure of the west coast of North America. An American geologist W. Hamilton, among others, had proposed a similar idea, and suspected that the sunken portion of the East Pacific Rise is now under the *Basin and Range Province,* causing the well-known peculiar features of that region. The Basin and Range Province covering much of Nevada, Arizona, and Utah exhibits high heat flow and extensional tectonics, just what we would expect in an area underlain by an active rift.

At this point McKenzie and Morgan's idea of an evolving triple junction came along. Their ingenious suggestion was soon elaborated in great detail by Tanya Atwater (1970), then a graduate student at Scripps Institution of Oceanography. Atwater made a careful analysis of the striped pattern of geomagnetic anomalies off the North American coast, tracing the history of how the oceanic ridge collided with the American continent and disappeared. As is apparent in Figure 3-10, both sides of the oceanic ridge (the East Pacific Rise) are visible in the south of the Gulf of California, whereas only the west side is visible in the north of the Gulf. If the ocean floor spreads symmetrically on both sides of the ridge, in accord with the sea-floor spreading hypothesis, the east side of the ridge must be under the American continent. In the model presented by Vine, McKenzie, Morgan, and Atwater, a continuous trench existed along the western United States until 30 million years ago. The new sea floor produced by the East Pacific Rise was constantly descending into this trench, generating volcanoes and deep-focus earthquakes. At present, however, this descent—and the volcanic and earthquake activity—has ceased, leaving behind such "fossils" as volcanic rocks.

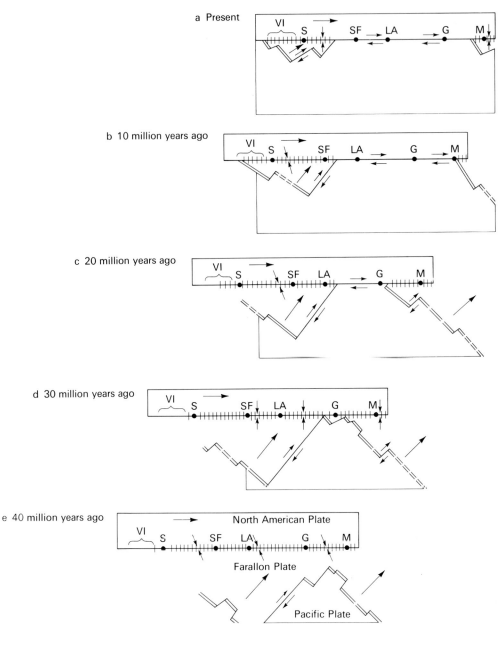

**FIGURE 4-9**
Schematic model of plate interactions, based on the assumption that the North American and Pacific plates moved with a constant relative motion of 6 centimeters per year parallel to the San Andreas fault. The coast is approximated as parallel to the San Andreas fault. Farallon-Pacific plate motions are approximated from the observed magnetic anomalies. The letter symbols refer to the following locations: VI—Vancouver Island; S—Seattle; SF—San Francisco; LA—Los Angeles; G—Guaymas; M—Mazatlán. The long arrows show motions of plates relative to the Pacific plate and the short arrows show relative motions along plate boundaries. [After T. Atwater, "Implications of Plate Tectonics for the Cenozoic Tectonic Evolution of Western North America," *Bull. GSA* **81**, p. 3513, 1970. Redrawn with permission of the author.]

Figure 4-9 represents one plausible model of the history of the interactions between the East Pacific Rise and the North American continent, as analyzed from the magnetic stripes. For those who find this model complicated, it might be helpful to refer to Figure 3-6 in which a similar concept is illustrated. (It may also be easier to understand if Figure 3-6 is turned sideways so that the San Andreas Fault extends horizontally.) Part (a) in this model represents the present period. Parts (b), (c), (d), and (e) show the progressively older periods. The Farallon Plate is the symmetrical counterpart to the main Pacific Plate, which formed on the east side of the East Pacific Rise. It vanished beneath the North American continent, leaving only a small segment behind on the east side of the Juan de Fuca and Gorda Ridges. The consumption of the Farallon Plate ceased when the East Pacific Rise and the North American continent collided, as shown in Figure 4-9. After the collision, the San Andreas Fault—a transform fault—began its activity on the west coast of North America. The initial contact between the East Pacific Rise and the North American continent is thought to have occurred about 30 million years ago (Figure 4-9d). From that time on, the two triple junctions (the T-F-F type in the north and the F-R-T type in the south) have been moving apart, thus lengthening the San Andreas Fault, and the relative movement between the two main plates now in contact has been parallel to the San Andreas Fault. This story provides a lucid, even revolutionary explanation for the geological history of western North America, and, at the same time, establishes the concept of the collision of ridge and trench even more firmly. The evidence, though difficult to accept, was overwhelming. Now we come to the question, what is a ridge?

## Oceanic Ridges

When the sea-floor spreading hypothesis was introduced, the postulate that an oceanic ridge was in fact a portion of the convection current welling up from within the mantle was rarely questioned. But this period of blissful ignorance did not last long. Once it was ascertained that oceanic ridges could migrate and that they either descended into the depth of the earth or were somehow annihilated, it also became apparent that they needed to be thoroughly re-examined in terms of the concept of the plate tectonics.

One of the ideas resulting from this re-examination was that oceanic ridges were simply cracks or "windows" in the plates that had been caused by their motions, and that these cracks were then

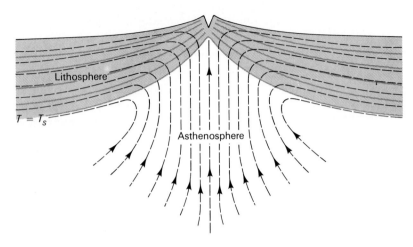

FIGURE 4-10
Schematic diagram of material flow (broken lines with arrows) and isotherms (solid lines) showing the thickening of the plate by cooling from above. The solidus temperature $T_s$ is the isotherm that represents the temperature between partially molten states: the solid lithosphere above is cooler than $T_s$, and the asthenosphere below is hotter.

filled by the ascending magma from the mantle. Then the oceanic ridges could migrate and they could eventually collide with oceanic trenches and be annihilated. Interpreted in a slightly different way, this idea could mean that oceanic ridges and the upwelling zone of mantle convection are not necessarily the same. Nonetheless, it is inevitable that hot material from the asthenosphere wells up at the oceanic ridges, since that is where the lithosphere is cracked and pulled apart. As the material approaches this opening, the pressure decreases so that partial melting accelerates. The basaltic magma then separates and cools, forming the crust of the sea floor. The plates then move farther away from one another. As the newest sea floor, which is the hottest and therefore mechanically the weakest, lies at the crests of the oceanic ridges, this continues to be the location of the cracking of the lithosphere. For this reason, the new material from below emerges in the ridge-crest area. This is why sea-floor spreading continues to be localized at the oceanic ridges. The oceanic crust is formed largely by intrusions of dikes of basaltic magma filling vertical cracks, but direct extrusion of magma onto the sea floor also occurs, as has been proved by dredging in oceanic ridge regions. Eruptions on the sea floor can be identified by the presence of lava with a unique morphology—called *pillow lava*—which is formed by rapid solidification owing to the presence of ocean water. Basalt magma quenched in this way acquires strong thermo-

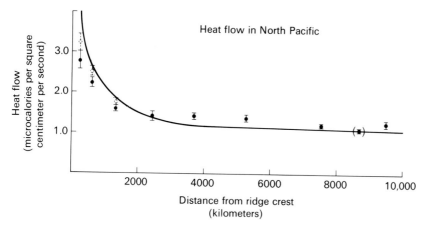

FIGURE 4-11

Comparison of observed heat-flow averages (represented by the symbol ♦ in the north Pacific with the theoretical profile (the curve) for a lithosphere 75 kilometers thick. Open symbols represent observed values increased by 15% to account for possible biasing effect near the ridge (see text). Note that the observed averages decrease with increasing distance from the ridge. [After J. G. Sclater and J. Francheteau, "The Implication of Terrestrial Heat Flow Observations on Current Tectonic and Geochemical Models of the Crust and Upper Mantle of the Earth." *Geophys. J.* **20**, p. 509, 1970. Redrawn with permission of the authors.]

remanent magnetism, which may produce the magnetic anomalies observed above the sea floor.

As the new sea floor spreads away from both sides of the ridge, it gradually cools from the surface downward and thickens. Surfaces within the lithosphere, or a plate, along which the temperature is constant can be represented by isotherms, as shown in Figure 4-10. The isotherm for the solidus temperature $T_s$ (the temperature at which melting begins) is especially important because rocks above the level of this isotherm have completely solidified as the lithosphere, whereas rocks below it are partially melted and constitute the asthenosphere. The actual value of $T_s$ depends on the mineralogical composition of the upper mantle and on whether water exists there or not. It is expected to vary from 1000°C to 1200°C for wet and dry mantle.

As mentioned previously, the heat flow from the sea floor is highest at the ridge crest and gradually decreases farther away from the crest. Earlier, scientists attributed this phenomenon to mantle convection, but the plate model we are discussing in this section seems to explain it more adequately: the decrease in heat flow is caused by a gradual cooling of the plate.

Figure 4-11 shows how the heat flow through the Pacific Ocean floor decreases with increasing distance from the ridge and hence with increasing age. The theoretical profile for the plate model, and agreement between it and the observations is quite satisfactory, except for the crestal zone of the ridge. Along ridge crests, many observations reveal unusually low heat flow as well as high heat flow, and this makes the average value of actual heat flow much lower than the theoretical one (page 55). C. Lister considers this discrepancy to be the result of a gigantic hydrothermal system that exists within the upper layers of the crust near the ridge crest. Within this system, heat is released into the ocean not only through the usual solid conduction through the rocks but also by the circulation of water through cracks in the rocks. Much vigorous investigation is being carried out to clarify the nature of hydrothermal circulation in the oceanic crust.

Another observation, by J. Sclater and J. Francheteau, provided the most important supportive evidence for the plate model. They noticed that the observed decrease in heat flow away from the ridge means that the isotherms are sloping downward away from the ridge, causing the plate to grow thicker (see Figure 4-10). The solidified portion of the plate is also gradually cooling and becoming denser, the corresponding decrease in volume being due to thermal contraction. As a result, the lithosphere sinks several kilometers deeper into the mantle and the depth of the ocean becomes greater over the colder and older parts of the sea floor. Sclater and others investigated the relationship between the depth of the oceans and the age of the sea floor and confirmed this reasoning. In Figure 4-12 the observed topography of the Pacific sea floor is compared with a theoretical model. The agreement is indeed remarkable. For example, it is a well known fact of oceanography that the eastern Pacific region is shallow and the western region deep, and yet it is a thrill to observe that the plate model can explain it as a universal and inevitable consequence of a simple phenomenon like thermal contraction.

The foregoing discussion summarizes the present plate tectonic model for the production of plates at oceanic ridges. The supportive details for these concepts will have to wait for future research. A. Miyashiro and his colleagues, for instance, have gathered a great many metamorphic rocks from the sea floor. The basaltic rocks that constitute the solidified crust must have been buried by some kind of mechanism to be metamorphosed and then exposed on the sea floor once again by actions such as fault movement. Another equally or even more important problem appears to be the actual geological, tectonic, or petrological processes that generate the ocean crust beneath the crest of mid-oceanic ridges. But to solve such a problem

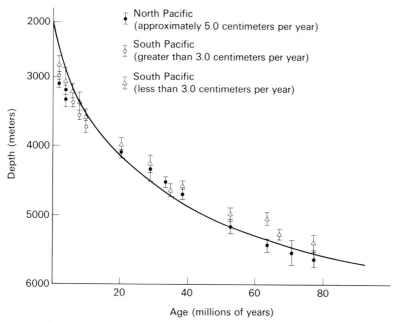

FIGURE 4-12
The average depth in the north and south Pacific plotted against the age of the oceanic crust. The symbols represent the various spreading rates indicated in the figure. The theoretical profile represented by the curve (proposed by Sclater and Francheteau) is for a lithosphere 100 kilometers thick. [After J. G. Sclater et al., "Elevation of Ridges and Elevation of the Central Eastern Pacific." *J. Geophys. Res.* **76,** p. 7888, 1971. Copyrighted by American Geophysical Union.]

one must employ techniques that have a resolving power much higher than is available in usual surface ship observation. A cooperative investigation by the United States and France—the FAMOUS Project (French American Mid-Ocean Undersea Study)—has been conducted recently in the crestal zone of the north Mid-Atlantic Ridge. Large numbers of research vessels and submersibles were employed in this joint goal of revealing what is actually happening at ocean ridge crests. Detailed heat-flow and magnetic measurements, rock samples, and high resolution photographs were obtained at the ridge crest itself and at nearby fracture zones. These showed that the process of spreading is much more complicated than previous observations from surface ships had led us to believe. In fact, in many places, the ridge was found to spread much faster on one side than on the other, and even at large angles to its axis.

## The Mesozoic Magnetic Stripes
## in the North Pacific

Ever since the well developed magnetic lineations in the northwestern corner of the Pacific off Japan and the Kuriles were discovered (see Figure 3-15), I had been attempting to correlate them with similar magnetic lineations in the eastern Pacific. At first it was thought that the lineations off Japan, being in the area farthest from the East Pacific Rise, might be those of the oldest crust produced by the Rise, and that this crust had migrated across the entire Pacific to be consumed in the Japan-Kurile trenches. But meanwhile scientists had discovered the Great Magnetic Bight (Figure 3-10) in the Gulf of Alaska and traced the datable Cenozoic anomaly lineations as far west as the eastern side of the Emperor Seamounts. Furthermore, the ages of the lineations in that area were believed to be younger toward the north—just the opposite of the order one would expect if the Japanese lineations were formed by the East Pacific Rise. The Japanese anomaly profiles did not correlate with those east of the Emperor Seamounts either. This discrepancy remained an enigma.

Early in 1972, when I was visiting the Lamont-Doherty Geological Observatory, I met R. Larson just as he was on the point of discovering a possible solution to the enigma. The prospect was most exciting to me because the problem had been on my mind for years. At that time, he and C. Chase had data that showed the existence of three sets of unidentified magnetic lineations, the Japanese, the Hawaiian, and the Phoenix, all in the Pacific (Figure 4-13). The crust of the Pacific Basin had been magnetically dated (see the shaded areas in Figure 3-10) back for a period of about 76 million years by the so-called Cenozoic magnetic isochrons. But beyond that there were no magnetic lineations! This vast area with no magnetic signature is called the magnetic quiet zone. The situation was the same in the Atlantic Ocean.

As discussed in Chapter 3, several explanations of the quiet zone had been proposed. Indeed, under such circumstances, theories vary greatly, depending on the philosophy of the scientist. Some scientists suspected that the lack of magnetic lineations indicated that the sea floor had been produced by a mechanism other than sea-floor spreading, and that the sea-floor spreading hypothesis itself did not apply to earlier periods of geological history. Others believed that the older sea floor had been produced by the sea-floor spreading mechanism, but that its magnetic lineations had subsequently been erased. The simplest theory was one that maintained that no reversals in the

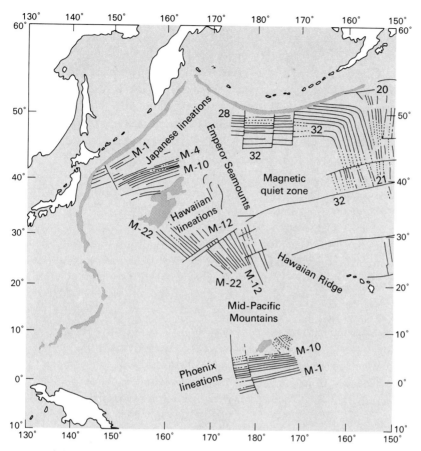

FIGURE 4-13
Mesozoic and Cenozoic magnetic lineations in the northwest Pacific. The
Mesozoic lineations are designated by M, with M-1 being the youngest. Numbers
without M are anomaly numbers for Cenozoic lineations (like those given for the
lineations in Figure 3-8). [After R. L. Larson and C. G. Chase, "Late Mesozoic
Evolution of the Western Pacific Ocean." *GSA Bull.* **83,** p. 3627, 1972.]

earth's magnetic field had taken place during the period in which that
particular portion of the sea floor was produced. While speculation
continued, a detailed picture of the three sets of unidentified lineations began to emerge from the Pacific areas, which were even farther
away from the ridge than the magnetic quiet zone.

Larson and Chase (1972) had confirmed that these three western
Pacific sets of lineations were indeed isochrons, and that they were
formed at the same period as a set of lineations in the western Atlantic outside the magnetic quiet zone. These Atlantic lineations,

discovered by P. Vogt, are called the Keathly set. What was more remarkable was that Larson and Chase, using the information from the Deep Sea Drilling Project, were able to determine the age of these lineations to be 110–150 million years old—back in the Mesozoic era. The lineations were named M-1, M-2, M-3, and so on, with M standing for Mesozoic and M-1 being the youngest (Figure 4-13). This work of Larson and Chase was an important breakthrough and their conclusions are fully incorporated in the compilations of the age of the ocean basin (Figures 3-10 and 3-14).

Thus the history of sea-floor spreading was traced back as far as the Mesozoic era, and a clue to the enigma of the north Pacific was found, although minute details are still unknown. At the same time, the history of the earth's magnetic field reversals was extended back from 76 million years to 150 million years—almost double the length of time. Compare Figure 4-14 with Figure 3-9. Figure 4-14 shows the long epoch of normal polarity between 85 million and 110 million years ago, which produced the Cretaceous quiet zone. It may be noticed that there is another long epoch of polarity in the Upper Jurassic, older than 148 million years. This corresponds to the very oldest part of the oceans and is called the Jurassic quiet zone. Also because the absence of geomagnetic reversals during these periods was known from earlier paleomagnetic studies of rocks found on land, the two lines of paleomagnetic research fit together beautifully.

## The Northward Movement of the Pacific Plate

Synthesizing all of the information discussed above, we come to the conclusion that sea-floor spreading has been going on since at least the time of formation of the oldest existing sea floor; it created first the Jurassic quiet zone, the Mesozoic sequence, then the Cretaceous quiet zone, and then the so-called Cenozoic magnetic sequence beginning with "anomaly 32" (see Figures 4-13 and 3-10).

The method Larson and Chase used to identify lineations that were far away from one another was a highly sophisticated technique developed by a Dutch scientist, H. Shouten. This method enables us to estimate even the magnetic latitude that a particular portion of the sea floor or the oceanic ridge possessed when that portion of sea floor was generated. This might imply that, by studying the magnetic lineations, paleomagnetism of the sea floor is possible. If we accept the postulate that the position of the geomagnetic pole, when averaged over a period of 10,000 years or so, coincides with the earth's

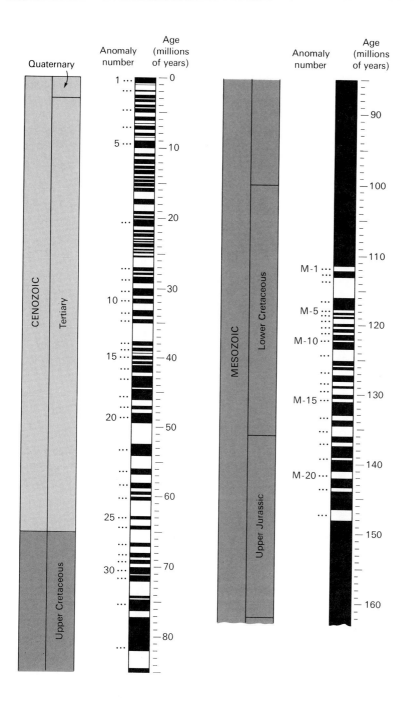

FIGURE 4-14
Geomagnetic reversal time scale from the present to the beginning of the Upper
Jurassic (162 million years ago). [After R. L. Larson and W. C. Pitman,
"Worldwide Correlation of Mesozoic Magnetic Anomalies and Its Implications."
*GSA Bull.* **83,** p. 3645, 1972.]

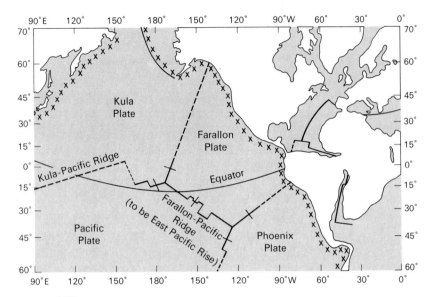

FIGURE 4-15
A possible plate configuration of the Pacific Ocean 110 million years ago. The crosses represent the subduction zones. [After R. L. Larson and W. C. Pitman, "Worldwide Correlation of Mesozoic Magnetic Anomalies and Its Implications." *GSA Bull.* **83**, p. 3645, 1972.]

rotational pole, then in principle we should be able to use paleomagnetism to determine the drift of oceanic plates, just as it was used to prove and measure continental drift. R. Larson and W. Pitman (1972) demonstrated the possible condition of the Pacific Ocean approximately 110 million years ago in Figure 4-15. According to their description, there were at least four plates, five oceanic ridges, and two R-R-R type triple junctions in the Pacific during that period. The Pacific Plate has been migrating to the north ever since, traveling 6000 kilometers or more and taking with it the oceanic ridges from which the sea floor continued to spread. The Farallon Plate was descending beneath the west coast of North America and the Kula Plate[*] was descending beneath the Aleutians at the same time that it was sinking beneath the Japan and Kurile Trenches. At present, both the Kula Plate and the Kula-Pacific Ridge have disappeared whereas the Pacific Plate, to the south of the Kula-Pacific Ridge, is in the process of underthrusting.

---

[*]The name *Kula* was proposed by J. A. Grow and T. Atwater in their paper on the Aleutian Arc (1970), Kula meaning "all gone" in an Athabaskan Indian dialect.

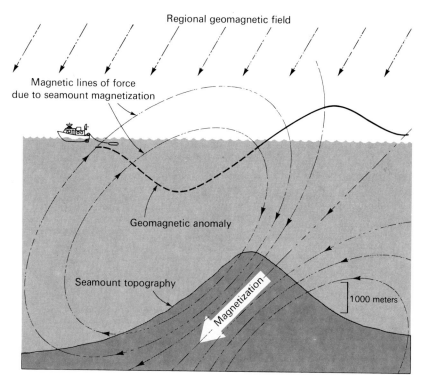

Regional geomagnetic field

Magnetic lines of force
due to seamount magnetization

Geomagnetic anomaly

Seamount topography

Magnetization

1000 meters

FIGURE 4-16
Schematic illustration of a magnetic cross section of a seamount. The geomagnetic
anomaly field produced by the magnetization of the seamount is represented by the
magnetic lines of force. The anomaly field is positive (solid curve) when it tends to
enhance the regional field at the sea surface, and negative (broken curve) when it
tends to cancel the regional field. Although only one cross section is shown here,
profiles of cross sections are used for computation.

V. Vacquier, I, and others have long been working on the paleomagnetism of the sea floor from a different approach. Many of the
innumerable seamounts that exist on the sea floor originated as
undersea volcanoes and are strongly magnetized. If we can determine
the direction and intensity of that magnetization, we should be able to
use such information in the same way we use paleomagnetism in
rocks on land. Yet, because seamounts lie thousands of meters deep
on the sea floor, it is practically impossible to gather oriented samples. Therefore, we sailed crisscross over the seamounts, surveying
the topography of the sea floor in great detail as well as the geomagnetic anomalies caused by the seamounts. The results of this
survey enabled us to determine by computer the direction and intensity of magnetization of the seamounts (Figure 4-16). By 1964 we had

estimated, on the basis of this research, that the seamounts scattered in the Pacific off the coast of Japan—which are considered to be Cretaceous in age—have migrated northward since their birth, spanning a distance of more than 6000 kilometers. This was fully supported by Larson and Chase's research, which they undertook in 1972, utilizing an entirely different method. We congratulated each other on the agreement of our results.

Since the Mesozoic era, nearly 10,000 kilometers of oceanic plate along with oceanic ridges seems to have descended beneath Japan. This is an important factor in the history of the Japanese island arc, as we shall see in the next chapter.

### A Possible Cause of Marine Transgression

We have already discussed how the rate of sea-floor spreading can be deduced from the width and the age of the magnetic lineations. Therefore, if we consider the Cretaceous magnetic quiet zone as an extremely wide stripe, the average spreading rate of the sea floor when such a magnetic quiet zone was being produced can also be inferred. Larson and Pitman did just that, and came up with an estimated spreading rate during this period—between 110 and 85 million years ago—which was about three times greater than that of other ages. They called it a *pulse* of sea-floor spreading and maintained that both sea-floor spreading and the subduction of oceanic plates into the trenches during this period was particularly intensive. They further suggested that this could have been the cause of mountain-building activity during that period, which was also believed to have been extensive.

Taking this idea a step fruther, in 1971 O. Hallam of England presented the theory that great marine transgressions could be attributed to pulses of rapid sea-floor spreading. Throughout the geological ages, there have been periods in which the land was deeply invaded by sea water. This phenomenon is called *transgression*. The opposite phenomenon in which the sea water retreats is called *regression*. These phenomena were well known in geology, but their cause has remained a mystery. We have discussed the condition of a newborn plate produced by sea-floor spreading: hot and elevated, it gradually cools down and shrinks, and the water above it gets deeper. If, during a pulse period, plates were produced several times faster than in normal periods, however, the sea floor would be generally more elevated because of the existence of wider areas of young hot plate that had not yet cooled off or shrunk. This would make the oceans of the world much shallower. Just as water would spill over the rim of a bowl if the bottom were raised, sea water would overflow and invade deeply into the land, causing transgression. When the pulse period was over, the sea water would retreat and regression occur. In 1973 J. Hays and W. Pitman applied Hallam's idea in a most ingenious way: they correlated a well known transgression that took place during the Cretaceous period with the rapid pulse in sea-floor spreading inferred from the magnetic lineations.

Such challenging attempts to elucidate geological riddles are truly exciting to observe; but not all scientists believe in ideas such as the pulse of sea-floor spreading. The skeptics question the validity of the spreading rates deduced from the widths of magnetic stripes, especially in the quiet zone. The width of the stripe may be measured accurately, but the times at which it started and ended are not yet quite established. This topic will be one of great interest in the immediate future of plate tectonics.

## Plate Tectonics and Orogenesis

By 1969, the concept of plate tectonics had already challenged the most basic problem of geology, namely orogenesis. Acceptance of plate tectonics would be much greater today if applied solely to activities that are presently taking place on the earth, or to a geometrical description of the sea-floor spreading process that has occurred in the past 80 million or even 150 million years—a reasonably easy period to trace. However, its application to geological activities of the earlier pre-Mesozoic period is considered by a number of geologists to be capricious at best. Two young geologists, John Dewey and John Bird, are the champions of its application to earlier periods. They are endeavoring to investigate the outermost limits of the implications inherent in plate tectonics, and maintain that the process of plate tectonics has been taking place for at least the past billion years of the earth's history. The theory of continental drift maintains that at one time the Atlantic did not exist. Dewey and Bird (1970) extended the possible course of events even further back in history and suggested that there was once an ocean that might be called the Proto Atlantic. In fact, such an idea had been previously raised by J. Tuzo Wilson and others.

If we believe the downthrusting of an oceanic plate into a trench is the origin of orogeny, and we then try to explain all of the Paleozoic orogenic zones—such as the Appalachians of North America and the Caledonian Mountains (which lie across Scotland and Scandinavia) in Europe—we need an ocean floor to be underthrust! So orogenic zones along which underthrusting has occurred are interpreted as evidence for the earlier existence of oceans. Looking for clues like this, Dewey, Bird, and their predecessors have thoroughly studied the results of the vast amount of precise geological research that has been done on the North Atlantic region, including Newfoundland, Greenland and Canada—data representing many years of scientific endeavor. They have concluded that the Atlantic must have once been open before it closed, and that subduction of the ocean floor took place at the trenches of the Proto Atlantic. This subduction was similar to what is happening today in the Pacific margin, causing deep earthquakes, igneous activity, metamorphism, and so forth. These activities created the Caledonian-Appalachian Mountain belt. The continued consumption of the oceanic plate of the Proto Atlantic finally caused the closure of the ocean to form part of Pangaea. Collision of the continents perhaps culminated in the formation of soaring mountain belts. Afterwards, when Pangaea started to break up again, the present Atlantic was created. When the new ocean reopened, however, the coastlines along it were not identical to that of the original Proto Atlantic. Instead the ancient orogenic belt on either side of the Proto-Atlantic was divided into complicated fragments. Dewey further presented a simple but daring model for orogenic cycles, arguing that since the surface dimension of the earth is constant, if one ocean, the Pacific for example, expands, the other, the Atlantic, will shrink. If the Atlantic expands, as it is doing at present, the dimension of the Pacific will decrease. Based on this hypothesis, an explanation of the long history of repeated orogenic cycles has been attempted in the manner demonstrated in Figure 4-17.

According to this idea, there are three types of oceans—the Atlantic, the Pacific, and the Mediterranean types. Both the Atlantic and the Pacific types are spreading at the ridges, but one is expanding and the other is contracting. The difference between them lies in the fact that subduction of the oceanic plate is going on in the Pacific type at the continental margin, whereas no subduction is taking place in the Atlantic type. The Mediterranean type is contracting and *not* spreading (Figure 4-17c and f). It has a subducting boundary but no ridge.

In this model, continental margins are classified into two types—the Atlantic type and the Pacific type—depending on whether sub-

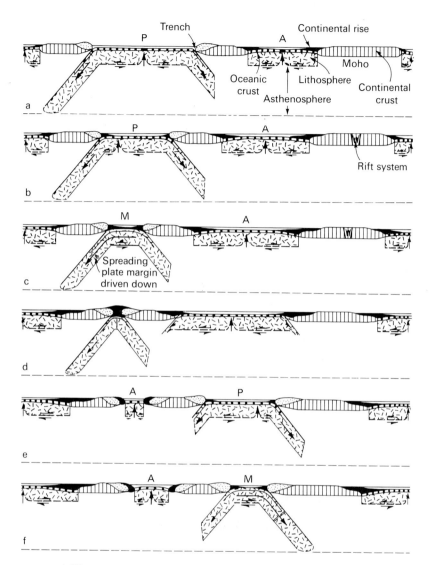

FIGURE 4-17
Schematic sections along a great circle of the globe showing the sequential development of the three main phases of ocean development: the Pacific type (P); the Atlantic type (A); the Mediterranean type (M). Parts (a), (b), and (c) show the spreading-expanding process that generates the A type phase in the ocean at the right. Simultaneously the ocean at the left, is undergoing the process of spreading-contracting—the P type phase, part (b)—and that of nonspreading-contracting—the M type phase, part (c). In (d) the ocean at the left has been closed, so that the expanding of the Atlantic phase at the right has to stop. In (e) the P type phase at the right and the A type phase at the left have started. If the ridge subducts (f), the ocean at the right becomes M type. [After J. F. Dewey, "Continental Margins: A Model for Conversion of Atlantic Type to Andean Type." *Earth Planet. Sci. Letters* **5,** p. 189, 1969.]

duction is occurring or not. The processes that are taking place in each type of continental margin are vastly different. In the Atlantic type, the sea is slowly subsiding as the plate cools and receives sediments from the continent, whereas in the Pacific type, all sorts of active orogenic events such as earthquakes and volcanic eruptions are occurring. For these reasons, the two types of continental margins are also classified as *passive* or *active* margins.

Dewey contended that once the process of oceanic development had gone through stages A to F in Figure 4-17 it began again. Thus oceans would evolve from the Atlantic type to the Pacific type and possibly through the Mediterranean type during any one orogenic cycle.

Even though Dewey and Bird are geologists themselves, their hypotheses are in direct opposition to traditional geological theories. Often a geologist instinctively avoids a simple generalization because he knows his area of specialization in such great detail that it is usually very easy for him to point out examples that cannot be explained by someone else's idealized theory. For this reason, Dewey and Bird's idea has not been well received by some geologists. Although it has not yet been verified as a scientific fact, it would seem worthwhile for geologists to try to find out whether this generalization is on the right track. To a geophysicist like me, the idea is fascinating. But the physical or chemical validity of a working hypothesis such as this can be verified only after the hypothesis itself has been clearly formulated. It will be difficult to examine the hypothesis from a theoretical point of view if it is expressed in a way that is vague, confused, or overly complex. Once the cause and effect of the phenomena are logically and clearly formulated, however, it becomes possible to examine its mechanism from a physical or mathematical point of view. Dewey and Bird have used their geological observations to suggest how events might have happened. But the problem of *why* things happened that way still seems to require further basic investigation. Why, for example, do igneous activity and metamorphism always occur when the cold oceanic plate subducts?

# Chapter 5

# Island Arcs

弧状列島

## The Journey's End

Along the northern and western margins of the Pacific lies a series of arcuate island chains—the Aleutians, the Kuriles, Japan, the Ryukyus, and the Philippines. The Izu-Bonin-Marianas arc branches off to the south from Japan. Further south lie Indonesia, the Solomons, the New Hebrides, the Tongas, and finally the Kermadecs. These are all island arcs. Their arcuate form is not their only common characteristic, however. They all have trenches, more than 6000 meters deep; most lie on the ocean side of each arc (see Figure 5-1). The west coast of the South American continent is not an island chain but may be included in this list of island arcs, or at least in the list of arcs with Pacific-type active margins as defined in the previous chapter.

These island arc and trench systems are of vital importance in the new theory of plate tectonics. If the mantle wells up and forms mid-oceanic ridges, and if the oceanic plates produced at the crest of such ridges spread horizontally, there must be zones at which the spreading oceanic plate descends again into the depths of the mantle. Otherwise, the surface area of the earth must increase with time. In fact, some scientists, such as S. W. Carey of Tasmania and B. Heezen have proposed such a model of an expanding earth. But the evidence overwhelmingly contradicts such a hypothesis. The earth is not "inflating." As an examination of a world map reveals, island arc and

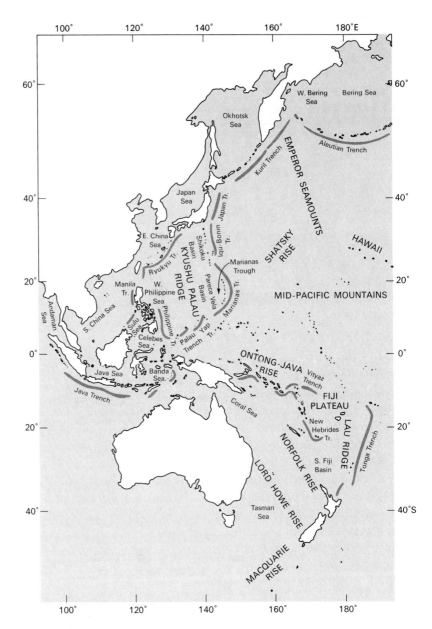

FIGURE 5-1
Island arcs in the western Pacific.

trench systems seem to be the most likely areas for the zones of subduction. A careful investigation of these systems is therefore critical to an understanding of the mobilist view of the earth.

The idea that island arc and trench systems are the journey's end for the ocean floor is not new. Although the theory of plate tectonics has been in existence only for several years, pioneering scientists such as A. Holmes, D. Griggs, and Vening Meinesz had already formulated a similar concept as early as the 1930s. Upon completing a famous experimental study of convection in the mantle, Griggs in 1939 pointed out that the geological features of island arcs could be explained if they were considered as developing on the downgoing zones of mantle flow. Meinesz reached a similar conclusion after measuring the gravity of oceanic regions in the 1930s. The Japanese earth scientists too—though traditionally reluctant to make sweeping hypotheses—accumulated a vast amount of data in their pioneering research, and the information they contributed was essential to the daring idea that was later to be expanded on such a grand scale.

## The Japanese Islands

The Japanese islands* are typical of the world's island arcs and have received the most extensive study. For these reasons, and also because of my own familiarity with this particular island arc, it will be the main topic of discussion in this chapter. Many of its characteristics are common to the other arcs as well, however.

Let us begin with the most basic factor, the topography. As is evident in Figure 5-2, the Japanese islands are arcuate in form. If judged by the coastlines of the islands, they appear to comprise four arcs—the Kurile arc, the Honshu arc, the Ryukyu arc, and the Izu-Bonin-Marianas arc. (Figure 5-1). However, if the contours of the submarine topography are taken into consideration, it becomes apparent that the above grouping is not entirely appropriate. "Ocean" and "land" are simply names attached to areas according to their present position relative to the level of the sea. Submarine topography, however, provides us with a much more significant way of perceiving the meaning of the existence of island arcs. A good example of this is the Japan Trench. As the topographical contours in Figure 5-2 show, the highest mountains of the Japanese islands are only 2000 meters above sea level, whereas the topography of the Kurile Trench, the Japan Trench, and the Izu-Bonin-Marianas Trench is much more pronounced, all three trenches being more than 6000 meters deep. The topography would seem to indicate that the Japanese islands actually comprise two island arc systems: one links the Kurile, the northeast Honshu, and the Izu-Bonin-Marianas

*This topic is discussed more extensively in a recent book by A. Sugimura and me (1973).

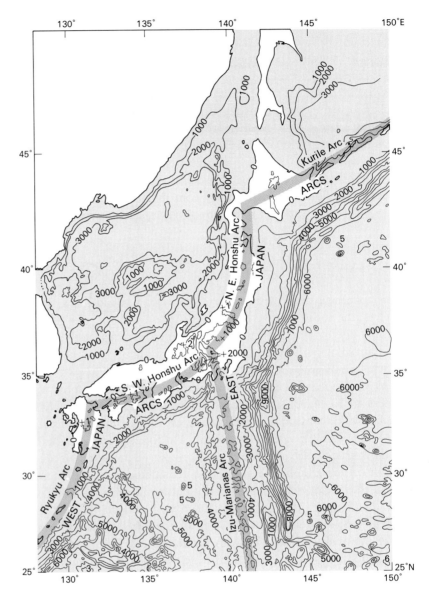

FIGURE 5-2
The topography of Japan and environs (as shown by 1000-meter contours) and the grouping of the island arc system. [After A. Sugimura and S. Uyeda, *Island Arcs: Japan and Its Environs*. Elsevier, 1973.]

Arcs (which we will call the *East Japan Arcs*), and the other links the Ryukyus, Kyushu, Shikoku and western Honshu Arcs (the *West Japan Arcs*). Thus, although Honshu looks like a single arc, it is now considered to be the place at which the two major arc systems meet.

A. Sugimura of the University of Tokyo was the first to conceive of grouping the Japanese islands into these two systems. Although the idea may seem slightly far-fetched, after we have considered all the geological and geophysical circumstances, it will be evident that it is quite valid.

One important feature island arcs have in common is a sea existing on the continent side of the chain of islands, such as the Sea of Japan, the Sea of Okhotsk, and the Philippine Sea (see Figure 5-1). These small seas are called *marginal seas* or, more specifically, *back-arc basins,* and their origin is considered to be critically related to the origin and development of island arcs.

## Gravity Anomalies

As mentioned earlier, the gravity measurements undertaken by Meinesz revealed the significance of the island arcs. Gravity is the force that pulls terrestrial bodies toward the center of the earth by universal gravitational attraction. (In order to simplify the topic at hand, we shall disregard the small centrifugal force generated by the earth's rotation.) If the earth were a perfectly homogeneous spherical body, gravity would be identical at any given spot on the earth. Actual measurement, however, reveals that force is variable—stronger in some places, weaker in others. The difference between the actual magnitude of the gravity measured and the *expected* magnitude is called the *gravity anomaly,* and it displays the irregularity of mass distribution in the earth's interior. Rocks that constitute continents are much heavier than sea water, and thus have stronger gravitational pull. Consequently, if the topography of the land and the ocean floor is known, one can calculate how much weaker the gravity is over the ocean than over the land. But the magnitude of gravitational attraction also depends on the distance between the mass and the spot at which the measurement is taken. It is a well known fact that the force of gravitation weakens as distance from the earth's center increases. Thus the strength of the gravity measured on top of a mountain, for example, is the result of two opposing factors—the greater pull of gravity due to the fact that the mountain is made of rocks rather than air, and the smaller pull of gravity due to the greater distance of the mountain top from the earth's center. The factors are reversed if a measurement is taken at the ocean surface. This means that, before we try to use gravity anomalies to learn something about the earth's interior, some adjustments must be made to the raw result of the measurements.

More interestingly, this condition does not apply to the southwest Honshu Arc, where the measured heat flow shows a pattern that is almost the reverse of that of the northeast Honshu Arc (Figure 5-16). Considering the long chain of inferences and the complex data processing employed in the assessment of temperatures from geomagnetic variation observations, the agreement with heat flow data is remarkable indeed.

## Seismic Waves and Temperature Distribution

Another way of estimating subterranean temperature is from the seismic wave velocity, another property that changes with temperature. Seismic waves slow down as they go through hot rocks. As the temperature approaches the melting point, the velocity drops even more drastically. At this point we must introduce another category of seismic waves—*surface waves*, which travel along the surface of the earth (as opposed to $P$ and $S$ waves, which travel *through* it).

A typical example of surface waves might be the ripples on the surface of a lake. Seismic surface waves include waves of various wave-lengths. Theoretically the propagation characteristics of surface waves are determined by the physical properties of the medium to the depth that is comparable to their wavelength. High-frequency surface waves (short wavelengths) tell us about the crust. Low-frequency waves (the longer wavelengths) tell us about the mantle. It is possible to estimate the physical properties of the entire upper mantle as a function of depth by investigating surface waves of various wavelengths. Using this technique, H. Kanamori and K. Abe have investigated the conditions of the upper mantle in the Japanese area. The upper mantle temperature distribution, as estimated from the results of their studies, is also in remarkable agreement with the temperature distribution as deduced from the measurements of heat flow (Figure 5-16). Kanamori and Abe also believe that the temperature is very high in the upper mantle beneath such areas as the Sea of Japan and eastern Philippine Sea, so high, in fact, that partial melting might be occurring at an unusually shallow depth of 30 or 40 kilometers.

Another important piece of information on the structure of the upper mantle of island arcs has been provided by seismology. It has long been observed in Japan that the distribution of the intensity of seismic vibrations is often very strange. For instance, when a deep earthquake occurs under the Sea of Japan at a depth of, say, 400

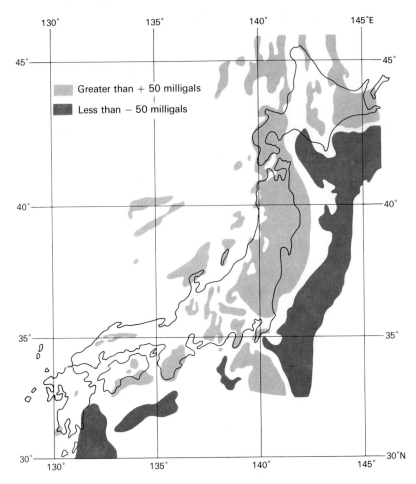

FIGURE 5-3
Free-air gravity anomaly in and around Japan. Positive gravity anomalies are
indicated in light gray, and negative anomalies in dark gray. (The milligal is a
convenient unit for measuring gravity anomalies. The earth's average gravity field
is about 980 gals.) [After Y. Tomoda, "Gravity Anomalies in the Pacific Ocean," in
P. J. Coleman, Ed., *The Western Pacific: Island Arcs, Marginal Seas,
Geochemistry.* University of Western Australia Press, 1973.]

First of all, the height of the position at which the gravity was
measured must be reduced to a standard level, usually to sea level.
This adjustment is called the height or *free-air* correction. After this
adjustment is made, standard gravity—the estimated gravity that
would be present if the earth's structure were completely uniform—is
subtracted. The difference is called the *free-air gravity anomaly*. Figure 5-3 illustrates the distribution of the free-air anomaly in and

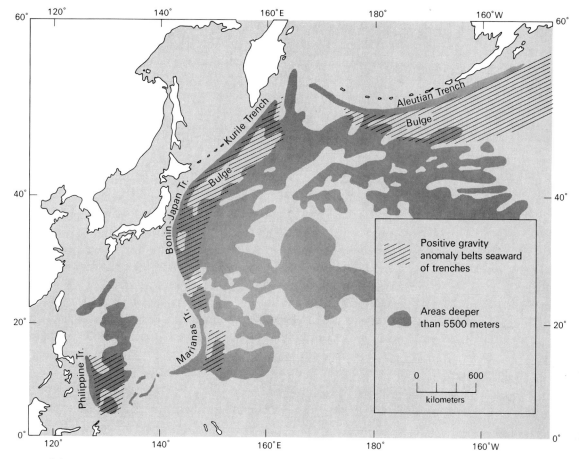

FIGURE 5-4
Positive gravity anomaly belts seaward of trenches, and general bathymetry of the northwest Pacific. Note that the positive anomalies are widest seaward of the central and eastern Aleutian Trench, and narrowest seaward of the Marianas Trench. [After A. B. Watts and M. Talwani, "Gravity Anomalies Seaward of Deep-Sea Trenches and their Tectonic Implications." *Geophys. J.* **36**, p. 57, 1974.]

around Japan. Notice that an extensive negative gravity anomaly belt exists in the Japan Trench area. The position of the axis of this belt is displaced slightly toward the continent from the axis of the trench. Part of the negative anomaly is due to the fact that the trench is filled with water instead of rock. However, after allowance is made for this condition, some part of the negative anomaly remains. This indicates that material of low density is present beneath the sea floor.

This result is amazing because it violates the principle of isostasy, which as we saw in Chapter 1, describes the observation that almost everywhere in the world the oceanic crust and the continental crust appear to be floating on the mantle like icebergs floating on water. The negative belt suggests that gravitational equilibrium is not being maintained in the trench. Unless some downward force prevailed here, the deep trench bottom would rise and equilibrium would be restored. The requirement that such a downward force be present can be easily understood if we imagine trying to force a floating log beneath the surface of the water. The log must be either pushed from above or pulled from below. Since it is obvious that no force exists to push from above the trenches, we are left with only one alternative—a force that is pulling down on the ocean floor from the depths of the earth. The theory of mantle convection currents explains this force as the drag created by the downward flow of the mantle. Belts of negative gravity anomaly like that in the Japan Trench are common to all the trenches of the world.

Recently, A. Watts and M. Talwani made an interesting study of the gravity anomalies seaward of the trenches. On the ocean side of many trenches, there are long topographic bulges several hundred meters high. Watts and Talwani found that the gravity anomaly at the bulges too are out of isostatic equilibrium. Again, applying the same logic as we did to the trench anomaly, we can see that the bulge must be supported by some force, an upward force in this case. Watts and Talwani ascribed the positive anomaly to the bending of the oceanic plate, which is strongly compressed as it approaches the trenches. Their interpretation seems reasonable, at least qualitatively, when we consider that the oceanic plate must force its way into the mantle at the trenches (see Figure 5-4 on the preceding page).

## Why Do Earthquakes Occur?

Another characteristic of island arcs is the seismicity. As we have seen in Figure 2-4, earthquakes do not occur just anywhere, but are concentrated in the circum-Pacific island arcs, the Alpine-Himalayan orogenic belts, and the mid-oceanic ridge systems. The detailed mechanisms aside, an earthquake is a phenomenon in which displacement occurs in the crust or in the upper mantle as a result of certain underground forces. The distribution of earthquake epicenters shown in Figure 2-4 would seem to suggest, therefore, that this process takes place mainly in the mid-oceanic ridge system, in the island arcs, and in the orogenic belts.

There are however, major differences between the earthquakes that occur in the island arc systems and those that occur in the mid-oceanic ridge systems: one is the depth of the foci.

**Deep Earthquakes.** Deep-focus earthquakes occur only in island arc areas.★ This fact is evident if we compare Figure 5-5 (which shows the distribution of the epicenters of deep-focus earthquakes) on the next page with Figure 2-4 (which shows the distribution of all earthquake epicenters). Now let us consider the distribution of deep earthquakes in more detail by examining that in the Japanese area more closely. Figure 5-6a on page 136 shows the distribution of the epicenters of the earthquakes in the area. Except for the few in the Japan Sea area, almost all of these earthquakes occurred at foci shallower than 60 kilometers. Figure 5-6b shows the epicenters of intermediate and deep-focus earthquakes.† As both parts of the figure show, even in island arc areas the major earthquake activities originate from shallow focal depths.

In this section, we will first examine the deep-focus activity. Note in Figure 5-6a, b that the focal depths are greater closer to the continent. The deepest one (near Siberia) is about 500 kilometers. This variation in the depth of the seismic foci is one of the remarkable features of island arc earthquakes; in comparison, those of the mid-oceanic ridges are all shallow.

As is demonstrated in Figure 5-6b, the foci of deep earthquakes seem to lie on a plane that inclines downward from the oceanic region toward the continent. The existence and distribution of deep-focus earthquakes was discovered in the 1930s by K. Wadati of Japan and his colleagues. Later H. Benioff of the United States investigated them further and stressed their significance, so that the inclined plane of foci has come to be known as the *Benioff zone*. Personally, I

---

★The Spanish deep-focus earthquake of March 29, 1954 (focal depth, 630 kilometers) is a notable exception.

†Earthquakes occurring at depths of more than 60 kilometers actually fall into two categories: intermediate-focus earthquakes, occurring at depths of 60 to 300 kilometers; deep-focus earthquakes, at depths of more than 300 kilometers. For the sake of simplicity, we will in some places refer to both types as *deep-focus*, where the distinction is not important.

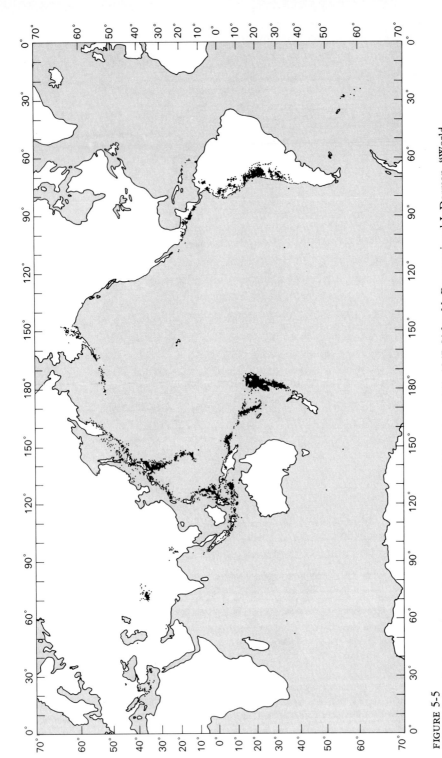

FIGURE 5-5
Distribution of epicenters of earthquakes deeper than 100 kilometers, 1961–1967. [After M. Barazangi and J. Dorman, "World Seismicity Map Compiled from ESSA Coast and Geodetic Survey Epicenter Data, 1961–1967." *Seismol. Soc. Amer. Bull.* **59**, p. 369, 1969.]

prefer to call it the *Wadati-Benioff zone* to give due credit to my fellow countryman.

What is the significance of this inclined plane of deep-focus earthquakes? First of all, in order for a fracture to occur that is sudden enough to generate seismic waves, the material must be brittle. Experimental evidence has shown that at high temperature and pressure, rocks tend to lose their brittleness, and flow rather than fracture. So the fact that earthquakes occur at depths where the mantle is hot and under high pressure is an enigma. But it is known that at low temperatures rocks tend to preserve their brittleness even under pressure: therefore, it would be a reasonable inference that along the Wadati-Benioff zone temperature must be unusually low. But it is obviously impossible for the thin, inclined zone to remain at a temperature lower than that of the surrounding mantle for a long time, because the hot mantle that surrounds it will soon heat it up. Apparently it can remain cool only if it is constantly supplied with new cold material in the form of the descending slab of lithosphere. This point was first stressed by D. McKenzie in 1969 in his plate tectonic model. McKenzie found that, if a cold plate of reasonable thickness (70 to 100 kilometers) plunges into the mantle at a speed of several centimeters per year, the center of the plate can remain cool down to 600 or 700 kilometers, the depth of the deepest earthquakes. In contrast, under the mid-oceanic ridges, where hot mantle material continues to well up, only shallow earthquakes occur because the rocks there are quite ductile below the shallow depths.

Thus the very occurrence of deep earthquakes is evidence for the subduction of cold plates. Moreover deep earthquakes would be expected to take place at the central part of the sinking plate in McKenzie's model. Prior to the introduction of this model, it was generally believed that deep earthquakes occurred at the interface between the sinking plate and the mantle surrounding it. The implications of this difference in theory are explained on page 140.

Now in addition to brittleness, there must also be strong stress if earthquakes are to occur. Information on the stress that exists at seismic foci can be obtained from research on earthquake mechanisms. By examining the direction of the initial motions of earthquake waves, the direction of the stress that caused the fracture at the foci can be estimated (see page 78). H. Honda of Japan and his colleagues had made pioneering contributions to the study of earthquake mechanisms as early as the 1930s. At that time many western seismologists believed that earthquakes were caused by a pair of equal and parallel forces acting in opposite directions termed a *force*

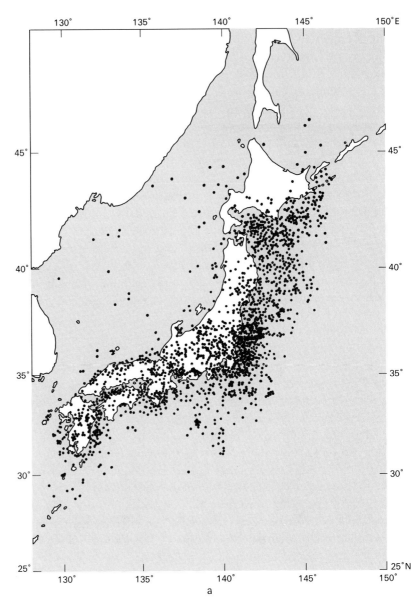

FIGURE 5-6
(a) Epicenters of earthquakes (of magnitude greater than 4) in the Japanese area,
1900–1950. The majority of these earthquakes were shallow-focus, except for those
occurring in the Sea of Japan area. (For an explanation of magnitude, see page 142
in the text.)

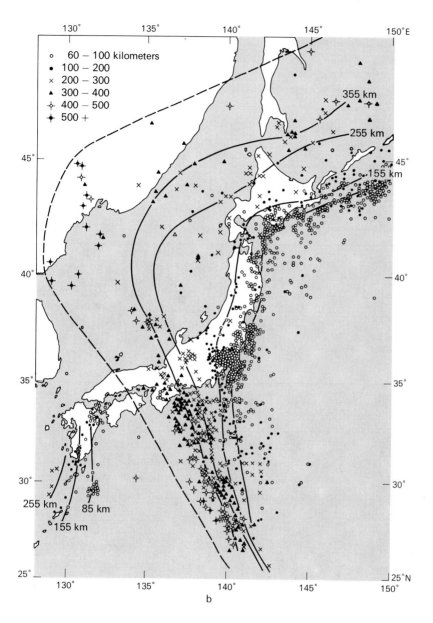

(b) Epicenters of intermediate and deep earthquakes, 1928–1962. [After Japan Meteorological Agency, 1958, 1966.]

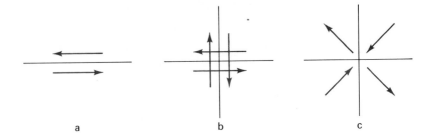

FIGURE 5-7
Proposed models of the forces operating at an earthquake focus.

*couple* (Figure 5-7a). But Honda and others showed that the cause was a *double couple* (Figure 5-7b)—that is, two force couples perpendicular to each other. The double couple in Figure 5-7b is mathematically equivalent to a pair of compressional and tensional forces as shown in Figure 5-7c. Theoretically, the double couple is more reasonable than the single couple, but it was only after a long period of controversy that Honda's concept was finally accepted. World recognition was delayed primarily because it was impossible to discriminate between the single-couple and double-couple models by observation of the first motion of $P$ waves alone. The validity of the double-couple model was finally confirmed by examining the distribution of the initial motion of $S$ waves. Detailed research on the initial motion of $S$ waves was difficult because, by the time they arrive, $P$ waves have already registered on the seismogram, thus making it hard to observe the arrival of the $S$ waves clearly. Japanese seismologists may have had an advantage in this research, because frequent earthquakes have prompted the installation of seismometers all over Japan, thus enabling them to arrive at the correct solution.

The directions of the two couples, as determined from the study of seismic waves, delineate the direction of the forces acting at the focus at the moment of the earthquake, that is, the earthquake-generating forces. Figure 5-8 demonstrates the distribution of these forces in the area around Japan. It shows the directions of the compressional stress of deep-focus earthquakes projected on a horizontal plane. It is clear from this figure that there is a regularity in the geographical distribution of such forces. The direction of the compressive axis is perpendicular to the direction of the arc. This again would seem to provide evidence of stress conditions consistent with the theory of plate tectonics.

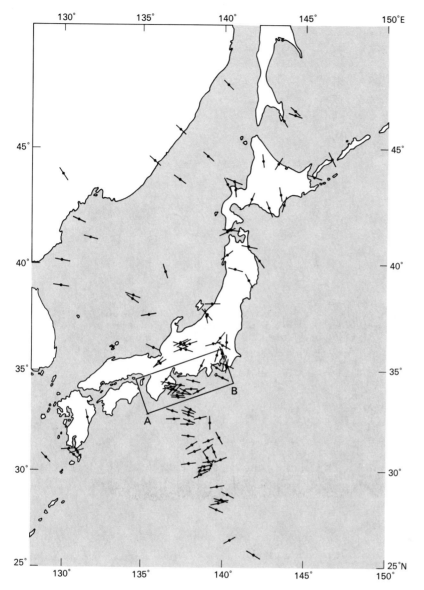

FIGURE 5-8
Directions of the maximum horizontal compressional stress for intermediate and deep earthquakes. The directions of the stress are indicated by the bars attached to each epicenter. For an explanation of the rectangle AB, see Figure 5-9. [After H. Honda, "On the Mechanism of Deep Earthquakes and the Stress in the Deep Layer of the Earth Crust," *Geophys. Mag.*, Japan Meteorological Agency, **8**, p. 179, 1934; M. Ichikawa, "Mechanism of the Earthquakes in and near Japan, 1950–1962," *Papers Meteorol. Geophys.* **16**, p. 201, 1966.]

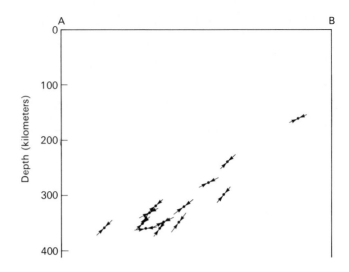

FIGURE 5-9
Directions of the principal compressional stresses for intermediate
and deep earthquakes in a vertical plane. Symbols are the
projections on the vertical plane passing AB in Figure 5-8. [Data
after H. Honda et al., "On the Mechanism of Earthquakes and the
Stresses Producing them in Japan and Its Vicinity." *Geophys.
Mag.*, Japan Meteorological Agency, **33**, p. 27, 1967.]

However the picture is not so simple when we look at the stresses
seen in a vertical cross section through the descending slab. Figure
5-9 shows the direction of compressional stress between A and B in
Figure 5-8 projected onto a vertical plane perpendicular to the trench.
We see that the direction of the compressional axes is parallel to the
plane of the descending slab as determined from seismology. What
does this mean? If the sea and continent (right and left in Figure 5-9)
are exerting stress on each other as they move together, wouldn't one
expect the direction of the compressive stress to be horizontal? Instead it is a *down dip* stress—that is, one parallel to the seismic plane
and directed down the direction of dip of the slab. Many people,
myself included, were troubled by this apparent dilemma for some
time. Then in 1969 B. Isacks and P. Molnar discovered that under
many other arcs the stresses are also down dip, but that sometimes it
was the tensional axis rather than the compressional axis that was
aligned parallel to the dip of the slab (Figure 5-10a).

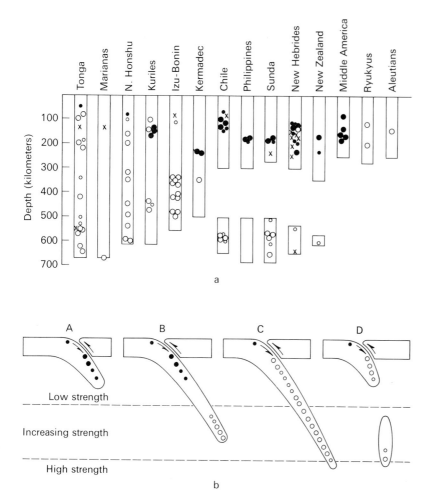

FIGURE 5-10

(a) Down-dip stress type plotted as a function of depth for fourteen regions. A filled circle represents down-dip extension; an unfilled circle represents down-dip compression; and the crosses represent orientations that satisfy neither of the preceding conditions. Smaller symbols represent stress types that have not been accurately determined. The enclosed rectangular areas indicate the approximate distribution of earthquakes as a function of depth by showing the maximum depth and the presence of gaps for the various zones.

(b) A diagram showing possible distribution of stresses in slabs of lithosphere that sink into the asthenosphere (A) and hit bottom (B and C). D represents a slab from which a piece has broken off. The symbols for the stress types are the same as in Figure 5-10a. In B and D gaps in the seismicity would be expected. Also shown are the underthrusting and the extensional stresses near the upper surface of the slabs due to the bending of the slab beneath the trenches. These features are inferred from the mechanisms of shallow earthquakes (to be explained in the text). [After B. Isacks and P. Molnar, "Mantle Earthquake Mechanisms and the Sinking of the Lithosphere," *Nature* **223**, p. 1121, 1969.]

Plate tectonics provided a simple, clear-cut solution to this problem too. Isacks and Molnar explained the phenomenon as follows: when the plate sinks, it meets resistance from the deeper, hard mantle as illustrated in Figure 5-10b. Moreover, the earthquakes occur in the middle of the plate or slab rather than at its interface with the mantle (see page 135). It would thus be more natural to describe the resistive force against the sinking plate as parallel to the slab. Let us note here that the mantle is soft in the asthenosphere but becomes harder at a depth of, say, 500 to 700 kilometers. As we will see in the next chapter, the descending slab is considered to sink by its own weight, so that there will be a tensional force in it if the resistance from outside is small. This explains the distributions of stresses shown in Figure 5-10b (at A, B, and D).

**Shallow Earthquakes.** Shallow earthquakes are much more spectacular because the energy released is so much greater and consequently their impact on human society is much more significant.

The size of an earthquake is usually represented by its magnitude $M$. We tend to think of earthquakes with great tremors as big and those with weak tremors as small; but this is not quite accurate, because no matter how big the earthquake might be, one cannot feel much shaking if the earthquake has occurred a great distance away. Just as the brightness of two lamps can be compared accurately by an observer only if the lamps are at an equal distance from the eye, the magnitudes of two earthquakes can be compared in terms of the amount of perceived shaking of the ground only if the observations are made at equal distances from the focus.

Earthquake magnitudes are generally determined by a standard method devised by C. Richter and B. Gutenberg. On the *Richter magnitude scale,* when the magnitude goes up by one unit, the seismic energy increases 30 times. The famous San Francisco earthquake of 1906 had a magnitude of 8.25; the great Kanto earthquake of Japan in 1923, a magnitude of 8.2; the Chilean earthquake of 1960, a magnitude of 8.3; and the Alaskan earthquake of 1964, a magnitude of 8.4. There is no record of an earthquake with a magnitude greater than 8.7. The seismic energy that accompanies an earthquake with a magnitude of 8 is approximately $10^{25}$ ergs*, which equals the energy of 10,000 atomic bombs of the type dropped on Hiroshima!

---

*The erg is the unit of energy: $10^7$ ergs is equivalent to 0.239 calorie.

Earthquakes with a magnitude of more than 7.5 are called *great earthquakes*. All of them are shallow, and nearly all of them occur in the circum-Pacific seismic belt. No great earthquakes have been known to occur in the mid-oceanic ridge areas. This is another major difference in seismicity between island arc systems and mid-oceanic ridge systems.

How do shallow earthquakes in island arc systems originate? From an analysis of the seismograms of great earthquakes and application of recent advancements in seismological theory, one can estimate not only the direction of the faulting that caused the earthquake, but also the size of displacement at the fault and even the size of the fault itself. H. Kanamori and others exhaustively investigated the great earthquakes that have occurred in the circum-Pacific zones. He proved that most of them took place when the sea floor of the Pacific was underthrust beneath the crust on the land side of the trench, and that the relative displacements along the faults were several meters.

In the meantime, S. M. Fedotov of the USSR, K. Mogi of Japan and L. Sykes demonstrated that areas of rupture caused by great historic earthquakes form oval patches that are roughly parallel to the arc. These patches tend to form a belt, but one that is discontinuous with numerous gaps at which no rupture has taken place during the time for which we have records. Furthermore, they showed that as more recent earthquakes occur, their focal areas tend to fill in these gaps rather than overlapping one another (Figure 5-11). They also pointed out that the time interval between great earthquakes in one area is in the neighborhood of 100 years.

One other fascinating fact should be mentioned here. It has long been known that great earthquakes in Japan cause the Japan Pacific coast to lift up by a few meters. But during an interquake period the coast gradually subsides.

If we put the foregoing information together, the following view is not too far-fetched: during the interquake period, the Pacific sea floor continues to underthrust several centimeters per year, dragging the crust on the land side down into the earth. When the deformation caused by this dragging reaches a critical point, a slip occurs at the boundary between the land crust and the oceanic crust and, as a result, the land side rebounds and is uplifted (see Figure 5-12). Assuming the average underthrust rate to be 5 centimeters per year and the interquake period to be 100 years, the slip would be 5 meters. This figure is in excellent agreement with the order of magnitude of the fault displacement—believed to be several meters—estimated from the seismic waves at the time of great earthquakes.

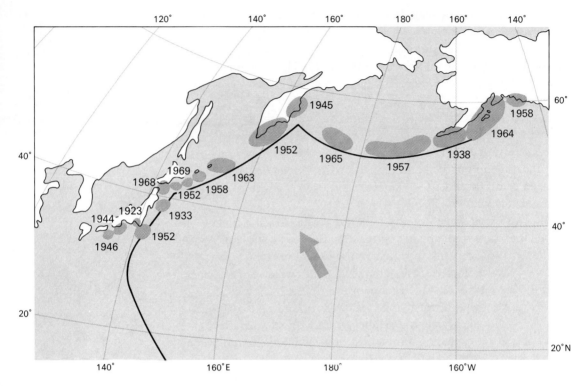

FIGURE 5-11
Distribution of focal regions as defined by after-shock areas of great earthquakes in the northwest Pacific margin. The year of the earthquake is indicated for each region. The arrow shows the approximate direction of motion of the Pacific Plate relative to the Eurasian and American Plates. [After K. Mogi, "Sequential Occurrences of Recent Great Earthquakes." *H. Phys. Earth* **16**, p. 30, 1968]

The fact that such fault areas are filling up the entire circum-Pacific zone also tends to support the plate tectonic theory, because it demonstrates that the entire rigid Pacific plate is subducting beneath the continental plate.

To summarize briefly the difference between shallow and deep earthquakes, the shallow earthquakes that occur in island arc regions are *inter*plate earthquakes that originate from the interaction of the oceanic and the landward plates—that is, the subduction of the first beneath the second; deep earthquakes, by comparison, can be termed *intra*plate earthquakes, because they occur within the subducting slab. The origin of earthquakes at other types of boundaries (diverging and transform fault boundaries) has already been discussed on pages 78–79.

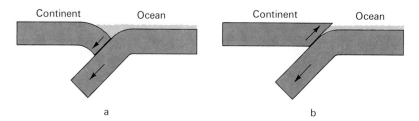

FIGURE 5-12
Movements assumed to take place under island arcs: (a) during the interquake
period; (b) at the time of an earthquake.

## The Underground Structure
## of the Japanese Islands

We have seen that the gravity anomalies seem to indicate that some
force besides gravity is pulling the trench floor into the earth in the
island arc regions. The testing of such an inference naturally requires
another method of observation—the determination of crustal structure through observation of seismic waves. Unusual layers of rock
underground can be detected by seismometers, because in such areas
the seismic waves will show irregular propagation. This procedure is
called seismic prospecting, and has long been used for the detection
of oil and other mineral resources. Not until after World War II,
however, was it widely used for the purely scientific purpose of studying the crustal structure to a depth of many tens of kilometers. In
1947 a vast quantity of German explosives were detonated on an
island in the Atlantic in order to dispose of them. This occasion—
presenting European seismologists with the opportunity to observe at
the same time, from different points on the continent, the seismic
waves generated from this explosion—turned out to be the world's
first large-scale experiment in explosion seismology.*

In Japan, a Research Group for Explosion Seismology was organized in 1950 and has been actively engaged since then in explor-
ing the crust and upper mantle. According to their studies, the crust
under the Japanese islands seems to consist of the following layers:

---

*As a result, for the first time they were able to obtain information on the deeper
structure of the European continent, since this kind of information can be gained
only by observing seismic waves from a distance: the further away the observer is
from the seismic wave source, the deeper his observations can be.

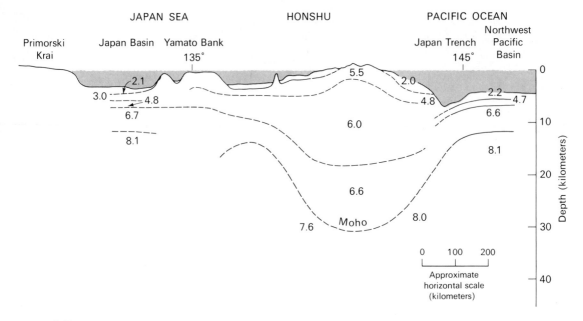

FIGURE 5-13
Crustal cross section across northeastern Japan. The numbers indicate *P*-wave velocity in kilometers per second. When there are breaks in the boundaries delineating the crustal layers, data were incomplete. [From S. Murauchi and M. Yasui, "Geophysical Investigations in the Seas around Japan." *Kagaku* 38, p. 196, 1968.]

an uppermost layer several kilometers thick with a *P* wave velocity of 5.5 kilometers per second, and a second layer 10 to 30 kilometers thick with a *P* wave velocity of 6.0 to 6.6 kilometers per second. This second layer may be composed of two distinct layers—one granitic and the other basaltic. The crust under the Japanese islands is thus almost continental. It was also found that the *P* wave velocity of the uppermost mantle (immediately below the Moho discontinuity) is about 7.6 to 8.0 kilometers per second. A velocity of 7.6 kilometers per second is significantly smaller than that of the world average (8.0 to 8.2 kilometers per second). These major results are illustrated in the crustal cross section of Figure 5-13. As can be seen in this figure, the crust under the Pacific Ocean is much thinner than the 30-kilometer crust under the Japanese islands. The crust under the Sea of Japan is also thin, especially in the region called the Japan Basin. There the sea is some 3500 meters deep and the crust is almost as thin as that of the Pacific Ocean. The area of the Yamato Bank, in the southwestern part of the Sea of Japan, has a thicker crust somewhat similar to the continental one.

Apparently the general tendency of the crust to become thin beneath the deep sea does not apply to trenches. Note in the figure that the crust under the Japan Trench is no thinner than that under the Pacific basin. This further verifies the inference derived from the results of gravity observation that isostatic equilibrium is not maintained in the oceanic trench areas.

## Volcanoes and the Belt of Volcanic Rocks

As has been observed so far, the various geophysical features of island arcs—such as gravity anomalies and earthquake activity—show a certain set of regularities. Such regularities exist not only in the Japanese area but in all island arcs. This fact would lead one to postulate that island arcs may be created by a certain mechanism common to all of them. The theory of plate tectonics suggests this mechanism is the subduction of the cold slab into the mantle.

But some geophysical features, despite their regularity, are not so easy to explain within the framework of the basic concept of subduction. For example, consider volcanic activity. The distribution of volcanoes is quite similar to that of earthquake foci. Volcanoes, like the foci, are concentrated on island arcs and oceanic ridges. There are active volcanoes in the ocean basins also, such as those of the Hawaiian Islands, but only a few. More than 90 percent of the world's active volcanoes are situated on the circum-Pacific island arcs. Figure 5-14 shows the distribution of volcanoes in Japan. As can clearly be observed, they are concentrated along the two systems of island arcs, the East Japan and the West Japan Arcs that are shown in Figure 5-2.

Although both volcanoes and earthquake foci exist in island arc systems, the distribution of each varies, as can be seen when the distribution of earthquakes in Figure 5-6a is compared to that of volcanoes in Figure 5-14. Whereas the majority of foci are concentrated on the oceanward sides of the island arcs, no volcanoes exist on the oceanward sides of the island arcs. The distribution of volcanoes and earthquakes may seem to be similar when indicated on a world map. Under more careful scrutiny, however, it becomes clear that their belts of concentration are displaced from one another. The oceanward boundaries of the volcanic zones can be clearly determined. These boundaries were named the *fronts of the volcanic belts*

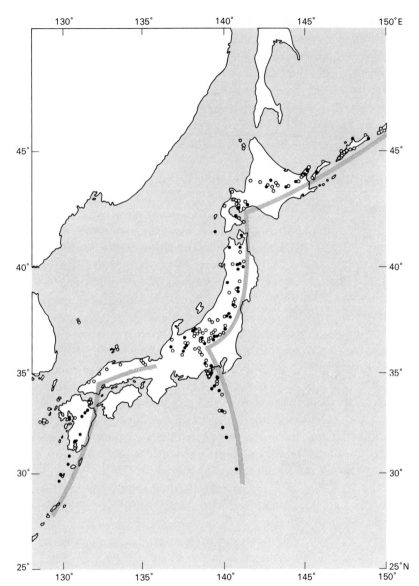

FIGURE 5-14
Distribution of volcanoes in Japan. Volcanic fronts are indicated by the gray shading. The closed circles indicate active volcanoes; the open circles, other Quaternary volcanoes. [After A. Sugimura and S. Uyeda, *Island Arcs: Japan and Its Environs.* Elsevier, 1973.]

or *volcanic fronts* by A. Sugimura in 1963. Why is it that not a single volcano exists on the oceanward side of the volcanic fronts?

Volcanic rocks are solidified lava that has been spewed out from volcanoes. The chemical composition and mineral assemblage of these rocks vary greatly, however, depending on the nature of the original magma—called the parental magma—at depth, and also on the various processes the magma has gone through before the eruption. For instance, if the temperature of the magma decreases, fractional crystallization takes place as minerals of certain composition crystallize and sink. Moreover, the remaining magma may react with the surrounding rocks. Both of these processes cause changes in the magma's composition. Modern petrology now enables us to estimate the chemical composition of the primary magma from the extruded lava, taking these complicated processes into consideration. Such estimates have revealed a regularity in the nature of volcanic rocks. The pioneering study in this field was conducted in Japan by T. Tomita in the 1930s, and more recently by the late H. Kuno. Their results showed a regularity in the chemical composition of primary magma. Among the various volcanic rocks, those exhibiting a chemical composition closest to the primary magma are basalts. For this reason, Kuno used basalt to determine the nature of parental magmas. Figure 5-15 shows the distribution of types of basaltic parental magmas in Japan, as proposed by Kuno. The magma that exists under the volcanoes near the volcanic front is called *tholeiite* basalt magma. The magma found farther from the volcanic front, and closer to the Asiatic continent, is called *alkaline* basalt magma, which is much higher in potassium and sodium content than tholeiite. Between these two regions of different types of magma lie basalts with high aluminum content. Kuno named such zones of magma *petrographic provinces.*

To explain the origin of these petrographic belts, which are parallel to the trench, Kuno proposed that the depth of magma production is shallower near the volcanic front and deeper near the continent. This hypothesis is based on findings from experimental petrology. H. S. Yoder of the Carnegie Institution and I. Kushiro of the University of Tokyo found that, the higher the pressure under which the magma is produced, the greater its content of potassium and sodium. The Geological Laboratory of the Carnegie Institution, with its excellent facilities for melting rocks under high pressure, is the mecca for experimental petrology. Since pressure is proportional to depth, the experimental finding of Yoder and Kushiro agrees, at least qualitatively, with Kuno's hypothesis—that the magma forms at greater depth near the continent. Kuno took the hypothesis one step further: he combined the depth of magma production with the Wadati-Benioff zone of deep-focus earthquakes, and postulated that magma production is related to the occurrence of deep earthquakes. At the time (in 1959), the concept of the subducting slab did not exist. Since its advent, a regularity in chemical composition of volcanic rocks has

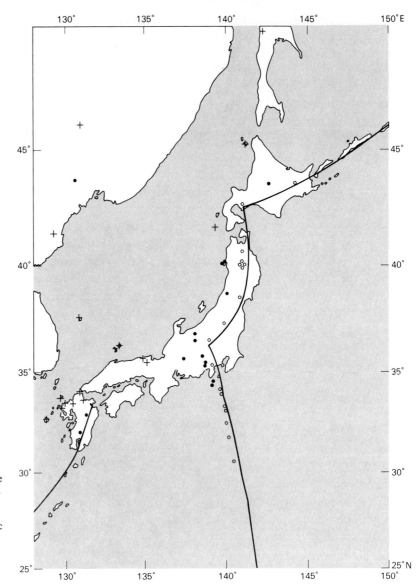

FIGURE 5-15
The distribution of types of
parental magmas in Japan and
environs: the open circles
designate tholeiite magma; the
closed circles, high-aluminum
basalt magma; the crosses,
alkaline basalt magma. The
solid line indicates the volcanic
front. [After H. Kuno,
"High-Alumina Basalt." *J.
Petrol.* **1**, p. 121, 1960.]

also been found in rocks other than basalts. Thus it became one of
the major problems of the new theory of plate tectonics to explain the
observed regularities in the chemical composition of volcanic magma
in the arcs. Many scientists have tried to synthesize the information
into a theory. A. E. Ringwood of Australia and his colleagues, and W.
Dickinson of the United States are among those who have made
significant contributions to the subject. We will not treat this prob-

lem in any more detail because an even more basic question requires our attention: how is magma produced under the island arcs in the first place?

## The Thermal State Beneath Island Arcs

Magma, which is produced in the depth of the earth's interior, most probably in the upper mantle, appears on the earth's surface as a result of volcanic eruptions. It is obvious that volcanic activity is largely determined by the thermal state of the upper mantle. Magma production occurs when the underground temperature rises, causing a localized melting. What is the reason for this rise in temperature and why is it localized in certain zones? Assuming that hot mantle material is extruded at the mid-oceanic ridges, it is reasonable to assume that magma production takes place there. Thus, the presence of volcanoes on the mid-oceanic ridges is readily explained. They are even more abundant in the circum-Pacific zones, however, where the cold sea floor is believed to be descending. It is hard to explain why. In order to do so, scientists realized that it would be necessary to understand the thermal state of the earth and, in particular, its peculiarities under island arcs. One observational approach to the problem was to make an extensive investigation of terrestrial heat flow.

We began the measurement of terrestrial heat flow in the Japan area in 1957. This problem was my postdoctoral research project at the Earthquake Research Institute, University of Tokyo. Although the project had to be started from scratch, it soon benefited from the generous cooperation of K. Horai, M. Yasui, and T. Watanabe. In addition, our oceanic heat flow work was greatly enhanced by joint projects with Scripps Institution of Oceanography and Lamont-Doherty Geological Observatory.*

In more recent years a number of Japanese scientists—I. Yokoyama, H. Mizutani, K. Baba, Y. Kono, and others—have been measuring the heat flow in various areas. Measurement of heat flow conducted in Korea with the cooperation of the Geological Survey of Korea was another successful international project. As a result of these various combined efforts, the heat flow of Japan and its environs has been most thoroughly studied. It is our current hope to extend these measurements into the main Asiatic continent and

*These projects were greatly benefited by funding under the United States-Japan Science Cooperation Program.

southeast Asia with the cooperation of scientists in the countries concerned.

The results of the observations on heat flow in the Japanese area up to 1970 are summarized in Figure 5-16. The regularity is remarkable. Terrestrial heat flow is low on the Pacific Ocean side of the island arcs or the trench areas, whereas that on the landward side is high: specifically, on the ocean side, where shallow earthquakes are frequent, the heat flow is less than 1 heat flow unit,* and on the landward side it is higher than 2 heat flow units. It should be mentioned that the seaward boundary of the high heat flow zone coincides rather well with the volcanic front shown in Figure 5-14. Most heat flow measurements are taken at stations that have been deliberately established away from the immediate vicinity of active volcanoes and hot springs. This way we can be sure that the higher heat flow registered on the landward side of island arcs is a regional state rather than a local one resulting from the proximity of individual volcanoes and hot springs. The findings demonstrate that the distribution of heat flow is closely and regionally related to the distribution of volcanoes and earthquakes.

Another notable feature in the distribution of heat flow (shown in Figure 5-16) is the fact that the high heat flow region extends toward the continent to include the marginal sea basins behind the arcs, such as the Sea of Okhotsk, the Sea of Japan, and the Okinawa Trough. The distribution of active volcanoes (Figure 5-14), however, shows a high concentration along the volcanic front and a decrease toward the continent. There are no active volcanoes in the Sea of Japan.

The regularity of heat flow distribution in island arc areas—low on the ocean side and high on the continent side or in marginal seas—applies to the Kurile Arc, the Northeast Honshu Arc, the Izu-Bonin-Marianas Arc and the Ryukyu Arc. But the Southwest Honshu Arc shows a high heat flow on the ocean side of the arc. Examination of Figures 5-6b and 5-14 will reveal that both deep earthquakes and active volcanoes are scarce in the area of the Southwest Honshu Arc; these observations support the view that it is not a typical active island arc. Further to the southwest, the Ryukyu Arc is more active and more typical.

How universal is this zonal distribution of high and low heat flow in the arcs of other parts of the world? Naturally we were eager to

---

*The world average value of terrestrial heat flow is about 1.5 heat flow units. (1 heat flow unit = 1/1,000,000 calorie per square centimeter per second.)

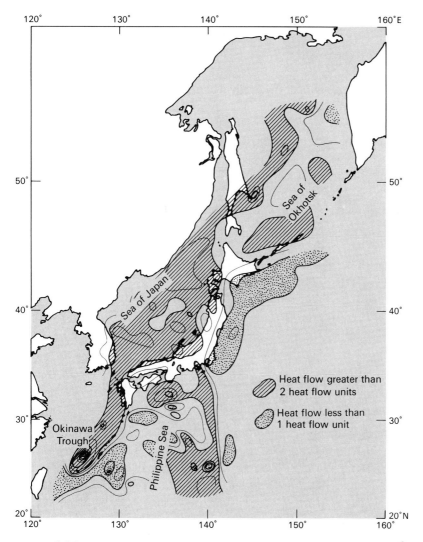

FIGURE 5-16
Heat flow distribution in and around Japan (in heat flow units). Contour interval is
0.5 heat flow unit.

compare the results of our observations with data obtained from other
arc areas. At that time, however, not many data on other arcs were
available. Thus an expedition was formed in 1969, to explore the arc
systems on the other side of the Pacific—along the west coast of
South America. T. Watanabe and I conducted heat flow measurements in the mines and oil fields of Ecuador, Bolivia, Peru, Chile,
and Argentina, with the help of local scientists and authorities, and of

mining geologists and scientists from various parts of the world who were working in these areas. The results of the measurements indicate the existence of a low heat flow zone seaward of the volcanic front, but the presence of a high heat flow zone landward of the front has not been verified. Certainly a more thorough survey remains to be carried out.

A number of surveys of other island arcs and back-arc marginal basins have been undertaken by various groups. So far measurements have been taken primarily in the oceanic areas around the arcs rather than on the islands themselves. The heat flow in the trench areas is almost always low, but that in the marginal seas behind the arcs is rather complex. In some, heat flow is high, and in others it is not. For instance, the North Fiji Basin (Fiji Plateau), the South China Sea, and the West Bering Sea exhibit high heat flow, whereas the South Fiji Basin, the West Philippine Basin and the East Bering Sea have either normal heat flow or heat flow of mixed high and low values. The heat flow in marginal basins seems to be closely related to the history of these basins, as we shall see later. Figure 5-17 is a schematic representation of heat flow in the Western Pacific area.

## The Distribution of Electrical Conductivity and the Thermal State of the Upper Mantle

Another possibly significant factor that reflects the thermal state of the earth's interior is the distribution of underground electrical conductivity. The rocks that compose the crust and the mantle are electrical semiconductors (so that they have some conductivity but are not good conductors), and their weak electrical conductivity increases with increasing temperature. Hence the distribution of electrical conductivity within the earth provides information about the temperature distribution.

The method of estimating the underground electrical conductivity is rather complicated. It is deduced from the time variations in the geomagnetic field. The main cause of geomagnetism is believed to be the electric currents flowing in the earth's metallic core. These currents are maintained by a kind of dynamo action in the core. A small part of the geomagnetic field, however, is induced by the presence of electric currents outside the earth. This externally induced geomagnetic field undergoes changes that are much more rapid than the very slow changes brought about by currents in the earth's core, such

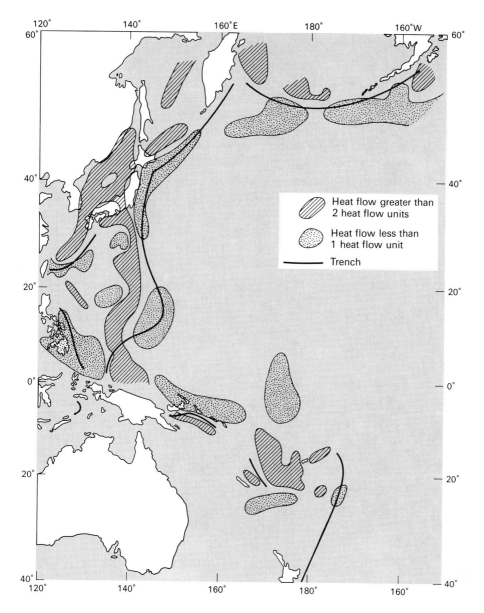

FIGURE 5-17
Simplified heat flow distribution in the western Pacific. [After T. Watanabe, "Heat Flow in the Western Margin of the Pacific," 1974. Redrawn with permission of the author.]

as the geomagnetic reversals (discussed in Chapter 3), whose duration is measured on a geological time scale. Time variations in the external field take place within hours or minutes, or even more rapidly. They are caused by changes in electrical currents in the upper atmosphere. These changes, in turn, are induced by changes in the flow of radiation and charged particles from the sun. It is well known, for example, that when a solar flare occurs, the geomagnetic field experiences especially large-scale disturbances called magnetic storms, but less dramatic changes also occur daily because the solar radiation at any given point on the earth changes daily.

Now, suppose that the magnetic field is changed by the electrical currents within the earth's upper atmosphere that are influenced by solar variations. The law of electromagnetic induction requires that a certain electric current will be induced in the earth's interior by this changing of magnetic field. This electric current will then produce a rapidly changing, though weak, secondary magnetic field. When we observe the geomagnetic field on the ground at the time of a magnetic storm, we measure, in effect, the sum total of the two kinds of variations. The first variation in the magnetic field is due to currents in the ionosphere; the second is due to the currents in the earth induced by the first. By isolating the secondary from the primary variations, one should theoretically be able to assess the strength of the electric current that has been induced in the earth. From this information one can further estimate the electrical conductivity of the earth. Estimation of underground electrical conductivity by this method was begun by the late S. Chapman of England and his coworkers more than 30 years ago. Then T. Rikitake and his colleagues discovered in 1955 that the mode of variations of the secondary geomagnetic field in the Japanese arc areas is quite anomalous compared to that in other areas of the world. From this they concluded that the electrical conductivity under the island arc of Japan was anomalous in comparison with other areas.

Anomalies in electrical conductivity suggest anomalies in temperature distribution, since electrical conductivity is an indirect indicator of the thermal state of a rock. According to subsequent detailed studies by Rikitake and his colleagues, the temperature distribution under Japan, as deduced from the study of the electrical conductivity anomaly, is in close agreement with that surmised from the distribution of terrestrial heat flow. This means that the temperature of the upper mantle underneath the Japanese islands, as deduced from the anomaly in electrical conductivity, is higher on the landward side and lower on the ocean side along the northeast Honshu Arc (Figure 5-2).

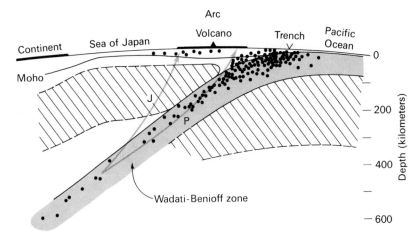

FIGURE 5-18
A model showing the various layers of the deep structure of the northern Japan arc. Dots represent earthquake foci. J is a typical path of seismic wave transmission from a deep-earthquake focus to the Sea of Japan side of the arc; P is the path of transmission from the same focus to the Pacific Ocean side. The light gray represents the cold layer in which seismic waves travel with high velocity and low absorption; the stripes represent the hot layer in which the waves travel with low velocity and high absorption. [After T. Utsu, "Seismological Evidence for Anomalous Structures of Island Arcs with Special Reference to the Japanese Region," *Review of Geophys. and Space Phys.* **9,** p. 839, 1971. Copyrighted by American Geophysical Union.]

kilometers, the seismic intensity is often stronger on the Pacific Ocean side than on the Sea of Japan side. Since the distance from the focus is greater on the ocean side, this phenomenon is an unexpected one that must be explained. M. Katsumata and T. Utsu of Japan interpreted this phenomenon as follows: typical paths of transmission of the seismic waves from a deep earthquake to the two sides of the arc are shown in Figure 5-18. If the rocks along path P are capable of transmitting seismic waves while absorbing much less than the rocks along path J, the Pacific side will feel more intense vibration than the Sea of Japan side. They concluded that the Japanese islands are underlaid by an inclined layer with low seismic absorption that intrudes from the Pacific side along the Wadati-Benioff zone. It was further noticed that seismic waves traveling through this layer of low absorption propagate at greater speed than those traveling through the upper mantle outside the layer. It is known that seismic velocity is greater and absorption is lower in cooler rocks. These observations clearly suggest the existence of a colder slab along the Wadati-Benioff zone (Figure 5-18). Thus the central concept of plate tectonics is again supported. A strikingly similar situation in the Tonga Arc of the South Pacific was found by J. Oliver and B. Isacks.

The thermal state of the upper mantle underneath island arc systems, then is quite complex. The most significant issue here is whether these thermal conditions support the concept of a descending slab or not. At present, it is commonly assumed that the subduction of the slab occurs because the slab is cold and, therefore, heavy. Given this assumption, such dynamic phenomena as shallow and deep-focus earthquakes, and such conditions as low heat flow, which signifies a low temperature and results in an absence of volcanoes, all seem explicable. Additional seismic evidence for the existence of the cold slab mentioned above lends support to the entire concept of the subduction of the oceanic plate. Yet, how can we explain such anomalous phenomena as high heat flow and high temperature within the upper mantle, and the existence of volcanoes on the landward side of island arcs? Does this not squarely conflict with the model in which the descending slab is considered to be cold?

## The Island Arc—An Orogenic Zone

Let us discuss briefly another observation before we explore the difficult problem of the conflict between the high temperatures in the upper mantle and the concept of the cold slab. One of the most significant features of the geological structure of the Japanese islands is the striking difference between northeast and southwest Japan. Northeast Japan exhibits pronounced features that are typical of active island arcs, such as a deep trench and a volcanic belt. This volcanic belt (see Figure 5-14) lies along the land side of northeast Japan and then turns to the south along the great geologic discontinuity called the *Fossa Magna* (see Figure 5-19) and extends into the Izu-Bonin-Marianas Arc. By contrast little present-day volcanic activity occurs in southwest Japan. This area is divided by the *Median Tectonic Line* shown in Figure 5-19. The geological structure there lies in belts that are generally parallel to the Median Tectonic Line. The most interesting feature of this structure is the distribution of regional metamorphic belts, also shown in the figure.

Metamorphic rocks are the secondary rocks that are formed when the original rocks undergo a transformation as a result of thermodynamic action. They can be roughly divided into two groups: those transformed by the presence of low temperature and high pressure, and those transformed by low pressure and high temperature. In

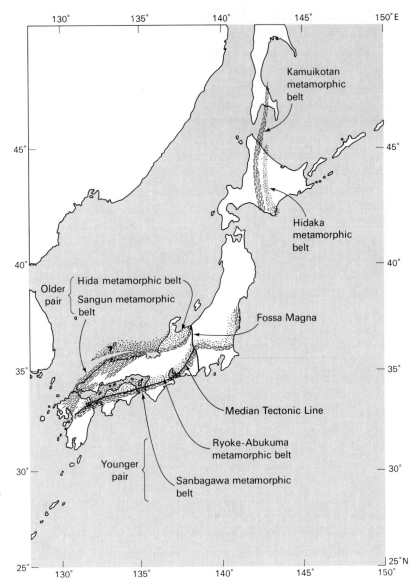

FIGURE 5-19
Regional metamorphic belts in
Japan. The Median Tectonic
Line and Fossa Magna are
shown in solid lines. [After A.
Miyashiro, "Evolution of
Metamorphic Belts." *J. Petrol.*
**2**, p. 277, 1961. Redrawn with
permission of the author.]

southwest Japan, these two kinds of metamorphic rocks are distributed with regularity, so that they form a pair of belts. The high-
pressure low-temperature metamorphic belt (the Sanbagawa Belt)
lies on the ocean side of the Median Tectonic Line while the low
pressure-high temperature belt (the Ryoke Belt) runs along the land
side of the Median Tectonic Line. These belts, then, are considered

to be indications of the type of underground activity that took place at the time of metamorphism. (The Ryoke-Sanbagawa Belts are the result of a metamorphism that took place in the Jurassic and Cretaceous periods.) Notice in Figure 5-19 that the metamorphic belts extend to the southern part of northeast Japan, beyond the Fossa Magna. This indicates that in Mesozoic or earlier times, Honshu acted as an orogenic belt. This is in contrast to the present configuration of northeast and southwest Japan. Another pair of metamorphic belts lie alongside the Sea of Japan in southwest Japan. The high-pressure low-temperature belt on the ocean side is also paralleled here by a low-pressure high-temperature belt on the land side. These are called the Hida and Sangun metamorphic belts. This pair of metamorphic belts is considered to have been formed in the Paleozoic age and hence it is older than the Ryoke-Sanbagawa pair to the south. The two pairs would indicate that the thermal state of the area in Paleozoic and Mesozoic times might have been much the same as that of an active zone today— that is, the temperature was low on the ocean side and high on the land side. Since southwest Japan is considered to be a rather inactive island arc at present, it can give us some idea of what the condition of active island arcs will be several hundred million years from now. Although a number of scientists must have recognized this possibility (at least subconsciously) for a long time, the first ones to postulate it—in the late 1950s and early '60s—were A. Miyashiro, A. Sugimura and T. Matsuda. Their theory was as follows: presently active island arcs are the very places at which orogenesis—the phenomenon that produced extensive mountain belts and metamorphism such as those seen in southwest Japan— is now taking place.

## Metamorphism and Heat Flow

What encouraged two of us—H. Takeuchi and me—to support the foregoing concept was the quantitative comparison illustrated in Figure 5-20. Line $A$ shows the distribution of underground temperature on the ocean side of the northeast Japanese arc, that is, the low heat-flow zone; line $B$ shows the temperature distribution on the land side, or the high heat-flow zone. These temperature distributions were deduced from the observation of heat flow by K. Horai and me in 1964. The horizontal axis in this figure shows the gradation of pressure and depth. The shaded area $a$ shows the temperature and pressure ranges that probably existed when the high-pressure low-

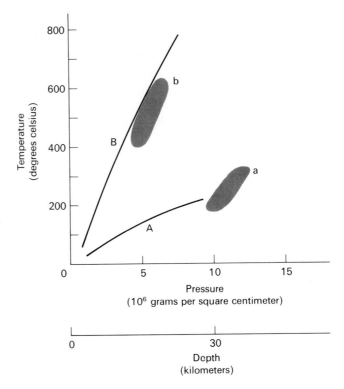

FIGURE 5-20
Temperature-depth relationships under present island arcs, and temperature-pressure ranges in metamorphic belts. Solid lines A and B represent the present temperature distribution in low heat flow and high heat flow zones. Shaded areas *a* and *b* are the temperature-pressure ranges for the high-pressure/low-temperature and low-pressure/high-temperature types of regional metamorphism. [After H. Takeuchi and S. Uyeda, "A Possibility of Present-Day Regional Metamorphism." *Tectonophysics* **2,** p. 59, 1965. Redrawn with permission of the authors.]

temperature metamorphism took place in the Sanbagawa Belt, and the shaded area *b* shows the probable ranges prevailing when metamorphism took place in the Ryoke Belt. These ranges of pressure and temperature were estimated by comparing the mineral assemblages found in the metamorphic rocks with the minerals produced in the laboratory under known conditions of pressure and temperature. Thus the temperature/pressure ranges were estimated by two separate methods—one based on heat flow along the presently active arc in northeast Japan and the other based on the composition of metamorphic rocks along the former orogenic zone in southwest Japan. Takeuchi and I thought that the agreement between them as shown in Figure 5-20 was too close to be coincidental, and suggested that the two types of regional metamorphism may be taking place now beneath the island arc of northeast Japan.

## Pacific-Type Orogeny

The similarity between our view about the activity of island arcs, which we consider to be orogenesis itself, and Dewey and Bird's revolutionary concept that all geological phenomena can be explained by plate tectonics (page 121) is readily apparent. But their view, though spectacular, is rather phenomenological. They seem to take it for granted that wherever the subduction of a plate takes place, earthquakes, volcanoes, and metamorphism will occur. The relationship between subduction and earthquakes is easy to accept, but the occurrence of volcanic activity and metamorphism (especially that of the low-pressure-high-temperature type) requires more explanation than they have given. Similarly, our postulation that the island arc represents a progressive stage in orogeny itself, demands more scrutiny of the physical mechanism involved. Again, the thermal aspect is especially difficult to explain.

In order to distinguish the orogenic process that creates great mountain ranges like the Himalayas from the process that takes place in island arcs, we will call the island arc process *Pacific-type* orogeny. The question of how melting and high temperature can occur at the same time that a cold plate is descending has long plagued the minds of scientists who are interested in the problems of "why."

## A Model for Pacific-Type Orogeny

Insofar as we adhere to the present commonly accepted model that a cold slab of lithosphere is descending under the arc, it is necessary to assume the existence of some kind of unique heating mechanism in order to explain Pacific-type orogeny. At present we are considering the possibility that frictional heat is the source of the thermal energy in Pacific-type orogeny. But how friction provides enough heat is not yet clear. Primitive man knew how to make fire by rubbing pieces of wood together, but modern science cannot determine how a cold plate, descending at the slow rate of several centimeters a year, could generate a sufficient amount of heat to cause the orogeny. In an effort to solve this problem, my colleagues (K. Hasebe and N. Fujii) and I, in 1970, set for ourselves the task of defining the physical mechanism that produces the heat. As a first step, we attempted to figure out quantitatively how much heat would be necessary in order to cause the observed high heat flow and volcanism on the continent side of island arcs. We found that numerical experiments with the use of electronic computers were the best approach. First of all, a normal oceanic cross section of the earth was assumed for the initial state. Then we created a model in which plate subduction began at a certain time. As the subduction continued, we assumed that a certain quantity of heat was produced along the interface between the sinking plate and the surrounding mantle. Then we calculated how the temperature distribution changed with the passage of time in the cross section. The heat capacity, the velocity of the descending plate, thermal conductivity, and the melting point of the materials were the factors included in the calculation. By changing the values of these quantities, we sought to determine numerically the conditions necessary to produce the observed heat flow distribution. The principal features of our model are shown in Figure 5-21. The dotted area denotes the descending cold plate.

An important element in calculations like ours is the time that has elapsed from the onset of the phenomenon to the present. Such a period can be estimated from geological observations, but not determined with precision. Since the geological structure of the Japanese islands could not have been formed within a span as short as 10 million years, and since it is known that the present island arcs came into being after the Mesozoic period, it is assumed that the period from the beginning of the underthrusting to the present has been 100 million years. Hence, in our efforts to explain Pacific-type orogeny our main objective was to determine what conditions 100 million years ago produced the present state. The amount of heat generated on the upper surface of the obliquely descending plate was estimated as follows: first, although the heat flow anomaly observed on the surface is about one heat flow unit,* 5 heat flow units must be generated on the upper surface of the slab. This large amount of heat is required because the cold plate subducts and most of the heat generated would be carried down with the cold sinking plate.

We also found that the value of thermal conductivity needed to transmit the heat to the earth's surface through the mantle would have to be about 10 times greater than that of mantle rocks. Thus our calculations indicated that a large amount of heat would have to be

---

*Recall that one heat flow unit equals $1/1,000,000$ calorie per square centimeter per second.

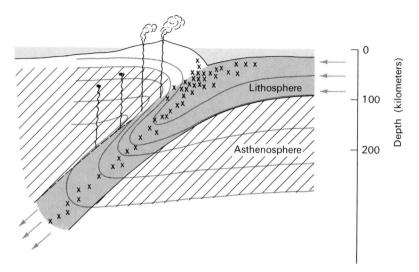

FIGURE 5-21
Model of a proposed heating mechanism under active island arcs that explains Pacific-type orogeny. The shaded area represents the descending cold plate (the lithosphere). Isotherms are represented by the curved lines. Arrows indicate the direction of movement, and the crosses indicate earthquakes. The wavy vertical lines indicate the ascent of magma. [After A. Sugimura and S. Uyeda, *Island Arcs: Japan and its Environs.* Elsevier, 1973.]

generated and that it would have to be conducted to the earth's surface very efficiently. At first such efficient heat transfer seemed impossible, but we found that it is possible if the heat is transferred not only by solid conduction but also by ascending magma. Quantitative examination, however, revealed that the lava actually found on the earth's surface is not enough—about ten times as much magma as the amount of existing lava would be necessary to transfer the heat. All these conditions are quite stringent and not very easily met. Some scientists believe that the high rate of frictional heating required by our model is not compatible with the assumed occurrence of melting needed for the transfer of heat, because melted rocks would act as a lubricant. This too is a major problem that remains to be solved!

## The Origin of the
## Sea of Japan—A Suggestion

If massive amounts of magma, 10 times greater than the amount of lava found on the surface, continued to ascend for as long as 100 million years, what could result? This question suggests a rather convenient hypothesis. The substance that rises to the surface during the 100-million-year period is equivalent to a column some 300 kilometers tall. If this material were piled on the earth's surface, a spectacular mountain 50 times higher than Everest would appear on the inner side of island arcs above the deep seismic zone. Since this has not happened, we should consider the possibility that the upper mantle now underlying the Sea of Japan has come up from underground during the last 100 million years. What does this mean? It is absurd to imagine that this area was once a depression 300 kilometers deep, but it may not be totally unreasonable to surmise that the present Japanese islands were once the eastern margin of the Asiatic Continent (Figure 5-22a) and subsequently migrated toward the Pacific Ocean to allow space for the ascending material (Figure 5-22b). This model would explain marginal seas—one of the features of island arcs—as the inevitable result of island arc tectonic activity. In our model the Pacific Plate is consumed as it descends into the inner side of the island arc, but some of it then reascends on the continental side to form the plate under the marginal sea. As marginal seas—such as the Sea of Okhotsk, the Sea of Japan, the Philippine Sea, and the East China Sea—increase in area, the arcuate form of the islands that outline these seas will become more and more pronounced. N. Kawai and his colleagues have found from paleomagnetic studies that the Honshu Arc has gradually bent throughout geological time. This bending would be the logical result of the process of marginal sea formation.

The origin of marginal seas like the Sea of Japan has long been the subject of controversy. It is difficult for a geologist to consider such a sea to be very old. Because rivers carry large amounts of earth and sand from the adjoining land, such a small sea would be filled in "no time at all" in geological terms (that is, within a mere 200 or 300 million years). Yet at present, the layer of sediment on the floor of the Sea of Japan is no more than 2 kilometers thick. This leads us to suspect that the Sea of Japan, as a sea, is fairly new.

An earlier theory was that until quite recently the Sea of Japan had been land and that the depression of this land had created the sea. This theory too met with a logical objection because, according to Archimedes' principle, land must have a thick crust beneath it to support its buoyancy (page 17). This means that if the land suddenly transformed into a sea, the ensuing depression would create an enormous gravity anomaly, and such an anomaly has not been observed above the Sea of Japan or anywhere else. This theory, then, could be viable only if the continental crust were *replaced* by an

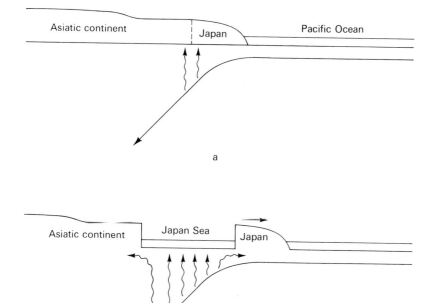

FIGURE 5-22
A model describing the way in which the Sea of Japan may have opened.

oceanic crust. Such "oceanization" of the continental crust has been
espoused by V. Beloussov, the Soviet scientist, in an attempt to explain marginal seas. However, most modern petrologists find ther-
modynamically unacceptable the idea that the rocks constituting the
continental crust could change into those making up the oceanic
crust, although the reverse process—conversion of oceanic crustal
material to continental material—is quite possible.

Given these considerations, I inferred that the Sea of Japan cannot
be regarded as a depressed piece of continent. In our efforts to determine which conditions did explain the present distribution of heat
flow and the mechanism of Pacific type orogeny (as described in the
preceding section) my coworkers and I suggested that frictional heating and the ascent of magma might have caused island arcs like Japan
to drift away from adjacent continents, creating marginal seas like the
Sea of Japan. Indeed the assertion that the Sea of Japan was formed
by the migration of the Japanese islands is not really new. As early as

the 1930s, T. Terada, a unique physicist as well as a natural philosopher at the University of Tokyo, reached the same conclusion in his study of the topography of the Sea of Japan. In recent years, two more of my countrymen, S. Murauchi and K. Nakamura have supported this hypothesis. I emphasize these other contributions because, although I have been preoccupied with my own line of reasoning about the Sea of Japan, I want to be sure not to neglect the valuable investigations conducted by other scientists on marginal seas.

Three major hypotheses on the origin of marginal seas have been put forth. We have discussed two—the *oceanization* and the *drifting of the island arcs*. The third is the *entrapment* hypothesis, which proposes that a new island arc is formed off shore in such a way that the part of the ocean landward of the arc is entrapped to become a marginal basin. This means that the marginal sea would have to be older than the island arc. Therefore, the hypothesis cannot be applied when the arc is very old, for the small basin would be quickly filled up with sediments from the surrounding land. Recently, A. Cooper, M. Marlow, and D. Scholl of the United States have postulated that the Bering Sea may be a trapped and ancient part of the Pacific Ocean. They claim to have discovered magnetic lineations in the Aleutian Basin that appear to be part of the M-sequence lineations (see page 116), which were produced as a part of the Kula Plate, later to be entrapped by the formation of the Aleutian Arc. Another sea that may have been formed by entrapment would be the western Philippine Basin.

No example of an "oceanized" marginal sea has been convincingly demonstrated as far as I know, although the proponents of the oceanization hypothesis consider almost all the marginal seas to be good examples.

Of the three hypotheses of marginal sea formation, drifting and possibly (for some seas) entrapment appear to be more realistic than oceanization. The verification of these hypotheses lies in the direct investigation of the marginal seas themselves. D. Karig (1974), among others, has contributed significantly to the problem of marginal sea formation. From his study of bottom topography, sediment distribution, and other geological factors, Karig in 1970 postulated that marginal seas were opened by extensional forces that pushed the arcs oceanward. He also introduced the concept that, in the course of island arc formation, the rise of material from the upper surface of the descending slab will split the arc longitudinally in two, and that an interarc basin is then formed between the two halves. These two

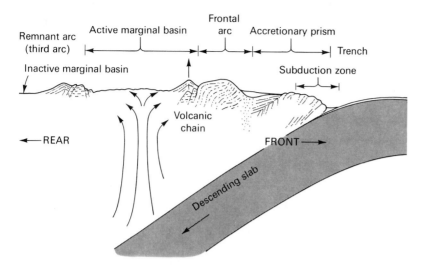

FIGURE 5-23
Model of a western Pacific island arc system, based on Karig's concept that an arc, in the course of its formation, is split into two parts—a frontal and a remnant arc—and that a marginal basin is thereby formed between them. The accretionary prism, on the landward wall of the trench, is a complex structure formed by ocean-floor sediment that has been scraped from the descending slab (it will be described in more detail in Chapter 6). [After D. E. Karig, "Evolution of Arc Systems in the Western Pacific." Reproduced with permission of *Ann. Rev. of Earth & Planet. Sci.*, Volume 2. Copyright © by Annual Reviews, Inc. All rights reserved.]

parts are called the *frontal* and *remnant* arcs. The frontal arc, having a volcanic chain just behind it, advances oceanward. For this reason the remnant arc is also called the *third* arc (Figure 5-23).

Karig was able to show that such a succession of events could explain certain features in various arc areas, including the Tonga-Kermadec, the New Hebrides, and the Marianas: he suggested that such extensional opening of marginal basins takes place episodically, producing a series of basins. Thus he contends that, within the Philippine Sea, the Marianas Trough, the Shikoku-Parece Vela Basins, and the West Philippine Basin are the successive results of increasingly older episodes of extensional opening (see Figure 5-1). As already mentioned, it is possible that the West Philippine Basin is an entrapped ocean. But the first two (the Marianas Trough and the Shikoku-Parece Vela Basins) may well be the results of two episodes, one presently active, and the other inactive now but formerly active in the Tertiary age. They are labeled as active and inactive marginal basins in Figure 5-23.

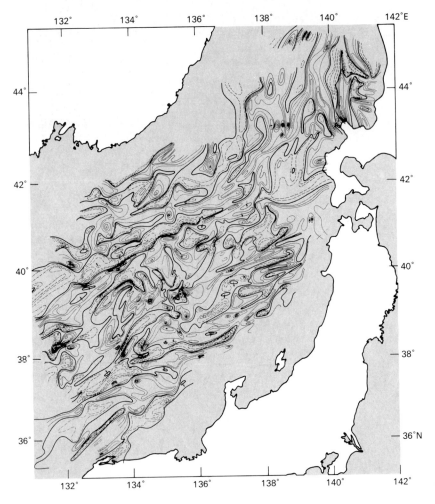

FIGURE 5-24
The magnetic anomalies of the Sea of Japan. Solid lines indicate the positive anomalies, and broken lines the negative anomalies. Although the anomalies are lineated in a northeast-southwest direction, the lineations are too irregular to permit their ages to be determined (compare with the lineations shown in Figure 2-8). The contour interval is 50 gammas. [After N. Isezaki and S. Uyeda, "Geomagnetic Anomaly Pattern in the Japan Sea." *Marine Geophys. Res.* **2**, p. 51, 1973.]

Recently, magnetic lineations have been mapped in the Shikoku Basin by both Japanese and Lamont scientists—Y. Tomoda and colleagues, and A. Watts and J. Weissel. Some of these lineations have been identified as Miocene in age, a finding that generally supports Karig's idea. It now appears that the formation of these marginal seas may be similar to sea-floor spreading (if not exactly identical), since they too have Vine-Matthews-Morley type magnetic lineations.

In the Sea of Japan, a magnetic survey was undertaken from two research vessels, the *Seifu-maru* of the Maizuru Marine Meteorological Observatory, and the *Kofu-maru* of the Hakodate Marine Meteorological Observatory. Then in 1970 we were able to conduct a more detailed survey from a fishing boat, the *83rd Daiei-maru*. Figure 5-24 shows the anomalies obtained from these surveys. The lineations seem to exist, but they are weak in intensity, discontinuous, and irregular in shape; therefore it is very difficult to identify their age by the standard magnetic reversal time scale. It is possible that even if the Sea of Japan did open as a consequence of the drifting of the Japanese arc, it was not by a typical sea-floor spreading that took place around a ridge, but by an irregular emplacement of magma—a *diffuse* spreading. However, N. Isezaki has suggested, on the basis of an intricate mathematical treatment, the existence of an axis of symmetry in the stripes of the Sea of Japan. If there is indeed such an axis, Vine-Matthews-Morley type spreading could have occurred here too. The problem is yet to be solved.

## Chapter 6

# The New View of the Earth

### New Global Tectonics

We have seen that the theory of continental drift underwent several stormy and controversial decades, during which it was almost completely discarded for a substantial period. It was revived by advances in paleomagnetism, strengthened by marine geophysics, and brought to maturity by the sea-floor spreading hypothesis, after which it rapidly evolved into what is now called plate tectonics.

We have also seen that there are many ways to study the earth. But if we are to move toward what might be called a "new view of the earth," we must synthesize the results obtained through the various approaches. Neither the sea-floor spreading hypothesis nor plate tectonics alone can fully support this new perspective that will eventually lead to an understanding of the way the earth continues to work; yet both are indispensable to such an understanding. In the new view, the earth is perceived as a mobile body, whereas in the traditional, opposing view it is regarded as immobile. Some describe this conflict as one between the mobilists (or drifters) and the fixists.

To summarize the concepts on which the mobilist view is founded, a continent is part of a lithospheric plate, and it moves with that plate. The lithosphere itself forms at the oceanic ridges. As soon as it forms by cooling, it begins to behave as a rigid plate. Several large and small lithospheric plates constitute the earth's surface. It is further postulated that all the large-scale phenomena occurring on the earth's surface at present can be attributed to the relative motions of the plates. When an oceanic plate collides with a continental plate, the oceanic plate descends, or subducts, beneath the continental plate, forming an oceanic trench. Subduction of an oceanic plate creates not only trenches, but also the island arcs. The activity of such an island arc epitomizes orogenesis itself. Where plates on which continents or island arcs rest collide with one another, a different kind of orogenesis occurs. Instead of subduction, continents and arcs form folded mountains. By this type of orogenesis, mountain ranges such as the Himalayas and the Alps are formed.

However, still to be answered is the basic question of what is causing these activities. Recall that it was Arthur Holmes who postulated in the 1920s that convection currents within the mantle cause the movement that carries continents, much as they would be carried along by a conveyor belt. This concept never died. It was kept alive, and provided a basis for the sea-floor spreading hypothesis proposed in the 1960s by Hess, Dietz, Wilson, and others. For the origin of island arcs and oceanic trenches, the late D. T. Griggs developed a theory in the 1930s, maintaining that they form where the flow of the mantle convection descends. This theory also survived to provide a framework for plate tectonics. At this point let us once again examine what is involved in "convection within the mantle."

## What Is Convection?

Everyone knows that a pot of boiling water shows a circulatory movement. When there is a disparity of density within a fluid, the heavier portion descends while the lighter portion rises to the surface. The convection that is observed within the pot is called thermal convection because it is caused by a density difference in the water created by a temperature difference. Thus if a potful of water is heated from below, the heated portion at the bottom expands, becomes light, and floats to the surface where it is cooled and becomes heavy, only to go down again. As it circulates in this way, the water is gradually heated throughout and gives off heat into the air. In short, thermal convection is a mode of transferring heat from the flame beneath the pot to the air above the pot. The convection that occurs within the mantle is also considered to be thermal convection. The deeper portion of the mantle is heated and expands, thus causing a circulatory flow.

Scientific investigation of such a simple and familiar phenomenon as the churning pot of water, however, was long neglected. H. Bénard of France was the first one to conduct the basic experimental research on thermal convection. The result of his experiment was published in 1906. In this experiment Bénard placed a thin film (between 0.5 and 1 millimeter) of paraffin on top of an iron cylinder and heated the cylinder from below. Bénard found that convection in the paraffin did not occur until the cylinder reached a certain temperature. He kept heating it past this point and observed that the heated portions began to ascend from everywhere, while the circumferential areas of such portions started to descend. After a while a regular hexagonal pattern appeared on the surface, as shown in Figure 6-1. The heated portion rises to the surface at the center of each hexagon in this pattern, and the cooled portion descends at the sides of the hexagon. Stirring cannot disrupt this regular pattern for long. In other words, the hexagonal pattern, called *Bénard's cells,* is stable. Bénard also discovered that the ratio between the thickness of the layer of liquid and the horizontal size of these regular hexagons—the length of their side, for instance—was close to one.

The theoretical study of Bénard's findings as a physical phenomenon, however, had to wait until 1916 when a paper by the famous Lord Rayleigh was published. He noted that heat from below does not necessarily cause convection as long as the heat can be conducted through the liquid to the air by normal thermal conduction. However, if there is too much heat to be transferred by thermal conduction, the heat accumulates at the bottom and the liquid there expands, becomes lighter, and starts to ascend. This is the beginning of convection. The viscosity of the liquid, however, tends to inhibit convection. In liquids with low viscosity, such as water, convection can occur readily, but in sticky liquids, such as oatmeal, convection is greatly hampered. This is the reason that oatmeal, for example, takes so long to cool.

Taking these factors into consideration, Rayleigh came up with the theoretical condition necessary for convection to occur. He maintained that when a nondimensional quantity defined as $R = \alpha\beta g h^4/k\eta$ reaches a certain number (about 1000 in actuality), thermal convection begins. If $R$ is less than this number, the heat is transmitted only by ordinary conductivity. In the expression for $R$, $h$ is the depth of the liquid layer; $\alpha$ is the coefficient of thermal expansion representing the fractional volume expansion caused by raising the temperature one degree; $\beta$ is the temperature gradient, that is, the rate at which temperature increases with depth within the layer; and $g$ rep-

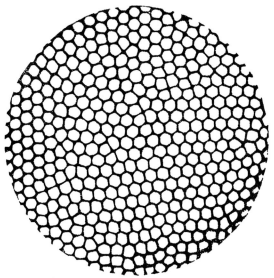

FIGURE 6-1
Bénard cells in paraffin. A drawing of one of Bénard's
original photographs. [After S. Chandrasekhar,
*Hydrodynamic and Hydromagnetic Stability.* The
Clarendon Press, 1961.]

resents the force of gravity. In the denominator, $k$ is the thermal
diffusivity, and $\eta$ is the viscosity. Qualitatively, the tendency for convection to occur as $R$ increases can be comprehended quite readily.
In the formula of $R$, the quantities that help convection are included
in the numerator, and those inhibiting it are included in the denominator. Rayleigh's theory thoroughly confirmed the results of
Bénard's experiments. The phenomena and theory described above
have been named Bénard-Rayleigh convection, and the quantity $R$ is
now called the *Rayleigh number.* The value of the Rayleigh number
at which convection starts is called the *critical Rayleigh number.*

It was found at the time of Bénard's experiment that the surfaces of
the sides of the hexagons, where the cooled portions descend, were
slightly more elevated than the surfaces of the centers, where the
heated portions ascend. This fact did not seem to conform to mantle
convection, because the ascending portions within the mantle rise to
form oceanic ridges while the descending portions form oceanic
trenches—exactly the opposite of Bénard's result. This riddle
haunted a large number of scientists until the 1950s when M. J.
Block of the United States and J. R. A. Pearson of England came up
with an answer. It was surface tension that had played a trick on

scientists. Astonishingly, according to Block and Pearson, the fluid motion observed in Bénard's experiment was caused not by thermal convection, but by the motion induced in the liquid by the changes of surface tension due to temperature variations. Since the importance of surface tension does decrease where the liquid layer is thick, Rayleigh's *theory* remains valid for thermal convection provided the layers are thick enough. The fact is, however, that Bénard's experiment, which Rayleigh had tried to theorize, had little bearing on the problem of thermal convection itself.

### Convection Within the Mantle

If we are to maintain that thermal convection occurs within the mantle, the applicability of Rayleigh's condition must first be examined. The first obstacle is the fact that the mantle is a solid body. Isn't a solid body one that never flows? However, consider an important fact that is not intuitively obvious. No matter how solid it is, no substance can permanently withstand the prolonged action of forces. A tall iron pillar, for instance, will not be able to support itself indefinitely but will bend and collapse in the course of a long period of time. Thus a solid body, even a crystalline substance like ice, actually does flow. It is simply that in most solids the rate of flow is imperceptibly slow. Can the earth's mantle too be fluid? As we saw in Chapter 1, the very fact that isostasy occurs means the mantle must have the property of flow: the great mountain ranges float on the mantle in accord with Archimedes' principle of buoyancy, and this principle is applicable only to fluids. The history of the Scandinavian peninsula is a good example of the fluid earth in action. Until about 10,000 years ago the peninsula was covered with a thick continental glacier and the land surface was depressed like a loaded raft. At the end of the last Ice Age the ice melted away, relieving the peninsula of the great weight of the ice sheet. This disturbed the isostatic equilibrium, and to regain equilibrium the Scandinavian peninsula started to rise and is still rising at a rate of several millimeters a year. The simplest explanation is that the mantle acts like a fluid with a viscosity of $10^{21}$ poises, a "poise" being the standard unit of viscosity. This phenomenon is called *postglacial rebound.*

Another indication of the earth's fluidity is its shape—an ellipsoid. The earth is not completely spherical because the centrifugal force of the earth's rotation causes it to bulge at the equator. This fact seems to prove that the earth as a whole can also act as a fluid body. It is thus an almost indisputable fact that in the course of long periods of time the earth's mantle flows in response to small but persistent forces. If we take all the relevant elements into consideration, the Rayleigh number for convection throughout the earth's entire mantle would be from $10^6$ to $10^8$, which is many orders of magnitude greater than the critical Rayleigh number (about 1000). Thus the earth's mantle fully satisfies Rayleigh's condition for convection.

With a viscosity value as great as $10^{21}$ in the denominator, it might seem that the Rayleigh number, $R$, would never have a large value. However, the depth of the entire mantle, $h$, which is included in the numerator to the fourth power, is so great that the Rayleigh number is very large, even with a large viscosity or a small thermal gradient (0.3°C per kilometer, for example). Insofar as one considers the mantle as a regular fluid (usually called a Newtonian fluid), the presence of convection seems natural and inevitable. It is not unlike the boiling in a kettle of water.

However, Rayleigh demonstrated only the condition necessary for the onset of convection under a set of idealized circumstances—not the kind of convection current that would actually occur under conditions in which the Rayleigh number was much greater than the critical value. A number of unresolved questions about mantle convection remain.

For one thing, the mantle—which we can regard in this discussion as a viscous fluid layer—exists as a spherical shell, rather than in the flat layer employed in Rayleigh's model. Moreover the heating of the liquid is not only from below, but the liquid mantle itself contains heat sources in the form of radioactivity. In order to understand this problem more realistically, there are many more factors to be considered. For instance, various physical properties, which were assumed constant in the Rayleigh theory, are actually functions of temperature and pressure. Viscosity, in particular, is known to be strongly dependent on temperature: it easily varies by many orders of magnitude with temperature variations of less than, say, 100°C.

More serious is the fact that we do not know for sure whether the flow property of the mantle is like that of ordinary fluids or Newtonian fluids. The definition of viscosity itself in the equations of motion then becomes ambiguous. Although perhaps not all of these factors should bother us, it is likely that some of them are very crucial. Certainly a formidable mathematical problem is posed: that of solving the equations of the motions of a deformable, rotating spherical shell whose viscosity may be non-Newtonian. We know this shell is highly stratified into a lithosphere, an asthenosphere, and a deeper mantle, and that it is being heated internally as well as from below. The worst of the difficulties is that the values of the important quantities —especially those concerned with viscosity—are so little known that realistic models are not easily identifiable, even if high-powered computing capabilities were available. Despite all these difficulties, however, an enormous amount of effort is being expended in attempts to advance our understanding of the convection within the mantle.

## Models and Reality

The famous Indian astronomer, S. Chandrasekhar of the University of Chicago, extended in 1949 the Rayleigh theory of thermal convection to spherical bodies, including spheres such as the earth that contain another spherical body (the core) inside. The results of his studies indicate that large-scale convection, with a flow encircling the whole earth, will occur only if the core is small. As the core grows, causing the thickness of the mantle to decrease, the pattern of convection cells undergoes abrupt changes, making the cell size smaller.

S. K. Runcorn welcomed Chandrasekhar's theory and used it to explain continental drift in 1965. He proposed that at one time the continents were floating on large convection cells. At certain stages in the earth's history, the growing core caused abrupt changes in the pattern of convection flow. These changes created instability in the position of the continents, causing them to split and drift apart (Figure 6-2). As proud as Runcorn was of his elegant thesis, it nevertheless met with a barrage of opposition. Some argued that since the actual state of the earth's interior was remote from the model to which the Rayleigh-Chandrasekhar theory was applied, and since the Rayleigh number within the mantle might be far greater than the value of the critical Rayleigh number, a simple pattern of convection flow, as postulated by Chandrasekhar, could not exist. Others maintained that a simple model was not at all applicable to the actual mantle in which viscosity varies. These were only a few of the many questions raised in response to this interesting idea.

A fundamental assumption has been that, given enough time, the earth can act as an ideal Newtonian fluid. Yet no one is absolutely sure that this is fact. We know little about what kind of flow properties the mantle has. We do not even know precisely what it consists of. The application of rheology (the science that investigates the

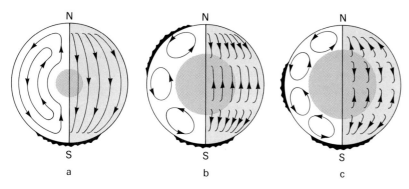

FIGURE 6-2
Convection in the mantle with different sizes of core. Parts (a) through (c) show the successive increases in the core size, the resulting changes in the pattern of convection flow, and the splitting of the continents. [Redrawn with permission of *Nature*.]

deformation and flow properties of matter) to the rocks of the mantle is still in its infancy. Although rocks that were once part of the mantle itself can be obtained, it is almost impossible to strain them at a rate slow enough to be relevant to geological phenomena. Despite this difficulty, rock deformation experiments have been inaugurated by D. Griggs and others, who are pioneers in the field of mantle rheology and convection. Even though these experiments are still far from conclusive, they have already suggested that the earth's mantle may have a property that is different from that of normal Newtonian fluids. In Newtonian fluids, the rate of strain or flow of the substance is proportional to the stress; in the earth's substance, however, the rate seems to increase exponentially by many powers proportionate to the stress. If so, flow in the mantle may be localized, as a jet current is. The flow will be slow in most areas, but once it is locally accelerated for some reason, its velocity will greatly increase. Although some investigators continue to insist the mantle behaves as a Newtonian fluid, the question raised by the rheology experiments—no matter how difficult —cannot be avoided and must be confronted by earth scientists today.

The difficulties are extremely complex. Various kinds of phase changes, like that between ice and water, as well as chemical changes, must occur within the mantle. When the ascending portion of the mantle reaches the surface through convection, a part of it probably separates to form the oceanic crust. At the same time, within the mantle itself, iron may be melting and sinking to the bottom, adding to the core. These are only a few of the many complex factors involved in mantle convection. It is not surprising, therefore, that an assumption by some—that the phenomena of sea-floor spreading and oceanic trench formation have been conclusively explained by a simple model—has been harshly criticized as too optimistic. This is only one example of the problem posed by the relationship between models and reality. Desirable though it may be to be able to explain a complex phenomenon by means of a simple model, reality may in fact be too complex: models are only models after all. Yet to reject any attempt at making simple models by flatly stating that one can never grasp even the physical principles of seemingly complex phenomena is self-defeating and unreasonable. Our present inability to quantify realistically the theory of mantle convection does not justify discounting it altogether. Such would no more reflect a scientific attitude than does undue optimism.

Those who recognize the possible importance of the layered nature of the mantle, especially the existence of the soft asthenosphere in the upper mantle, tend to maintain that the convection current does not run through the whole mantle but is confined to the asthenosphere. Such a stand, though reasonable, raises another problem. The Rayleigh theory and other more sophisticated theories all require that the *aspect ratio*—that is, the ratio of the horizontal scale of convection cells to the vertical scale—be close to one. Experiments also support this conclusion. The cells, then, in order to exist in the mantle should also have a horizontal scale comparable to the thickness of the asthenosphere—a few hundred kilometers.

However, the movements observed on the earth's surface—continental drift and sea-floor spreading—occur across a much larger horizontal scale, on the order of many thousands of kilometers. This disparity between the two horizontal lengths has bothered many scientists. At this point, however, H. Takeuchi and M. Sakata of Japan showed that if one devises a model of the whole mantle as composed of a low viscosity asthenosphere and a high viscosity lower mantle, the flow pattern, at least at the onset of convection, should be like that shown in Figure 6-3. Here, the flow is concentrated in the upper soft layer and the return flow is distributed deep into the underlying high viscosity layer. This means that the horizontal scale of the whole cell can be much larger than the depth of the asthenosphere. This result seems to explain to some extent the disparity between the two hori-

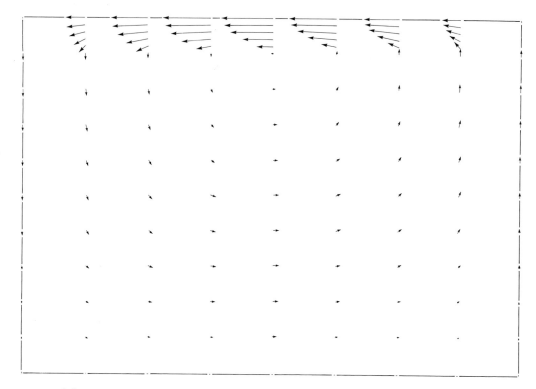

FIGURE 6-3
A possible flow pattern in which viscosity increases with depth. [After H. Takeuchi and M. Sakata, "Convection in a Mantle with Variable Viscosity." *J. Geophys. Res.* **75**, p. 921, 1970. Copyrighted by American Geophysical Union.]

zontal scales. However, I continue to feel that the problem of the aspect ratio of convection cells has not been solved. We will come back to this topic in another section.

## Hypothesis Versus Fact

In a field such as earth science, in which repeated demonstrations of phenomena with the use of artificial experiments is intrinsically difficult, the conflict of model versus fact is acute. Scientists often discuss whether to believe or disbelieve in hypotheses such as sea-floor spreading and plate tectonics, but belief is not really the issue. These are "working hypotheses," and the first condition to be fulfilled by a hypothesis is that it explain one or more phenomena in such a way that our understanding is increased. Whether a hypothesis is scientifically sound or not must be evaluated in terms of other standards in addition to its workability. Obviously we cannot "accept" just any hypothesis, but if a working hypothesis were to be immediately rejected unless it could be completely verified, what would be the point of formulating it to begin with? A useful hypothesis will become stronger and more profound, the more stringently it is checked against reality. Often this process occurs as follows: one adopts a working hypothesis, which prompts him to predict something as a logical consequence of the hypothesis, and the prediction is then tested by independent observations or experiments. At some point in this process, insurmountable difficulties may be encountered. The hypothesis will then have to be abandoned. Nevertheless, by the time the hypothesis is abandoned, our knowledge will be more advanced than it was before the hypothesis existed. This is the nature of a hypothesis: and when we present one, we are in fact saying, "We are not stating this as a fact. We are only proposing that a phenomenon be considered in this way, and we are continuing in our attempt to examine the workability of this approach." Is there any other way?

## The Thermal History of the Earth: An Example of a Developing Hypothesis

Conclusive answers to questions on the origin of the earth may never be attained. Only a little over thirty years ago, people believed the primordial earth was a ball of fire that had spun off from the sun, and the history of the earth was thought to be that of a cooling process. It was in the 19th century that scientists—led by such forerunners as the famous Lord Kelvin—had studied the thermal history of the earth on the basis of the assumption that the earth had evolved from a hot mush to its present cool hardened state. Then, at the turn of the century, A. Becquerel and M. Curie discovered radioactivity and the radioactive elements that exist in the earth's interior. This made scientists realize that even if the earth is cooling, it also contains a heat source within it. Soon they were busy recalculating the thermal history of the earth to take this heat source into account. Even without the heat source, the earth is so large that it would not cool readily. The presence of a heat source would further inhibit the cooling process. Calculation showed that, if the earth containing heat sources was originally hot, it could not have cooled to the present state! Thus, the concept of the earth as a once hot ball of fire that had gradually cooled was challenged. In the 1940s, the astronomer's concept of the origin of the sun and the solar system underwent a considerable change: some even proposed an entirely different idea. The earth could not have originated in the sun, they insisted; rather, it is an accumulation of cool cosmic dust.

At approximately the same time, a number of findings were added to our knowledge of the thermal conductivity of the earth's interior—another critical factor in the study of the earth's thermal history. Thermal conductivity had formerly been treated as a constant in a simple thermal conduction equation. However, it was found to decrease as temperature increased, meaning the earth is even more difficult to cool than was previously suspected. By the end of the 1950s, however, scientists came up with the prediction that thermal conductivity would increase once the temperature passed beyond about 1000 to 1200°C because heat can be transmitted by radiation, even in such substances as rocks. Simple theoretical considerations by S. P. Clark, based on the law of thermal radiation, indicated that thermal conductivity by radiation would increase in proportion to the cube of the absolute temperature. In essence, at high temperature rocks are nearly transparent to heat. Numerous calculations incorporating this radiative conductivity have been made on the thermal history of the earth.

Because of the many factors that had to be considered, the equations for the earth's thermal history became increasingly complex until they defied almost any attempt to solve them. What saved the day was none other than the advent of high-speed computers at the end of the 1950s. Scientists such as E. A. Lubimova of the USSR and G. J. F. MacDonald of the United States calculated, with the aid of numerous models, what the conditions of the earth might have been in the beginning in order to end up in its present state, taking all the known factors into consideration. The earth at present has a large-scale layered structure of core and mantle. Any theory of the earth's history must explain this primary feature. To form such a structure, total melting of the earth early in its history was assumed to be the most likely explanation. This argument was one of the strongest foundations of the hot-origin hypothesis.

The studies of Lubimova and MacDonald, however, demonstrated that even if the earth had originated in a cool condition 3 to 4½ billion years ago, there would still have been sufficient heat to melt the mantle, causing iron to accumulate in the center to form the core, and lighter elements to rise to the surface to form the crust. The cold-origin hypothesis was greatly supported by this demonstration. But the models these scientists used for their calculations were all solid bodies, so that the only mode of heat transfer they considered was conduction and radiation in a solid body.* Although they spoke of the large-scale motions within the earth that would be needed to form the core-mantle structure, their calculations were based strictly on solid state conduction theory until the melting point was reached, and sometimes computations were made in the same way for the temperature range beyond the melting point!

If mantle convection on a grand scale has been taking place continually, both below and above the melting point—as we maintain in this book—it would be meaningless to calculate the earth's past condition without taking into consideration heat transmission by convection. If convection occurs within the earth, the earth would have cooled much more rapidly, just as clear soup cools much faster than thick oatmeal. D. Tozer of England convincingly pointed out in 1969, that the earth starts to convect at a certain temperature below the melting point, so that the uppermost temperature in the earth is effectively controlled by this softening temperature, regardless of the initial conditions. So far, however, a thorough study of the thermal history of the earth from this perspective has yet to be completed.

If indeed the earth was formed by gathering cold cosmic dust that was gradually warmed, the heavy iron particles melting and gathering in the center to form the core, an enormous amount of gravitational energy must have been released as heat. F. Birch, an authority on solid earth geophysics, was one of the first to point out that the energy released by the settling iron would be great enough to heat the whole earth by about 1000°C and to melt it all at once. In that case, calculations in which such a great heat source was not taken into consideration would be pointless. Recently, D. Anderson suggested that as cosmic gas condensed and started to accrete, the core materials accreted first, thereby forming the layered structure of the core and mantle of the earth at the time of its formation. This suggestion was based on the cosmological consideration that the iron forming the

---

*In the late 1960s, however, it was discovered by Y. Fukao, H. Mizutani, and me (all of Japan), and by J. Aronson and R. McConnell of the United States, that the importance of the radiative contribution to thermal conductivity in earth-forming materials is much smaller than that assumed in these calculations.

185
The Thermal History of the
Earth: An Example of a
Developing Hypothesis

core was one of the earlier condensates of cosmic gas. If so, one need not account for the gravitational energy of iron settling. The resolution of the problem, however, depends on a better understanding of the process by which the earth and planets were in fact formed. Additional important factors require more cosmogonic insight. If the earth is an accretion of cosmic dust that was originally scattered throughout space and then accumulated to form a ball—the earth— an enormous amount of gravitational energy must have been transformed into heat as this happened. This energy is considered to be orders of magnitude greater than the energy released during the separation of the core from the mantle! If the accretion of the earth was a rapid process, the heat would have been effectively trapped in the earth; if it was slow, the heat would be radiated into space. Thus, depending on how rapidly the earth grew, it is possible that very high temperature may have prevailed during the earth's formation.

Radioactive heat sources, as is well known, decay with time. This means that more heat was generated by radioactivity in the past. Some radioisotopes, such as aluminum-26 ($^{26}$Al), iron-60 ($^{60}$Fe), and chlorine-36 ($^{36}$Cl), decay so rapidly that they are now extinct. But at the time matter was synthesized in the universe, it is estimated that a great quantity of these short-lived radioisotopes were also born. If the birth of the earth was not much later than, say, $10^7$ years after nucleosynthesis, those isotopes could also have contributed greatly to the earth's heat.

Generally, it was thought that the accretion of the earth took some $10^8$ years or more. If so, neither heat source—the gravitational energy of accretion or the short-lived radioactivity—was very influential in the process. But recently several lines of evidence from geochemistry and cosmochemistry, as well as theories of cosmogony, have indicated that the accretion process could have been much more rapid. H. Mizutani and T. Matsui of Japan have suggested it might have taken about $10^3$ years! If they are right, the earth in the beginning must have had a tremendous amount of heat. Thus Tozer's softening temperature would have been easily attained, meaning that the earth has been steadily convecting ever since to expel the heat. The formation of the core would have been rapidly completed during this process. Such a possibility would seem to revive the century-old hot-origin hypothesis. But the logic behind it is entirely different.

One of the reasons the hot-origin hypothesis was rejected was the fact that the earth, as a gigantic solid body, would be hard to cool. If active mantle convection is assumed, however, such an obstacle can be overcome. Furthermore, the idea that the earth's internal heat source would inevitably create a condition of extremely high temperature seems dubious when mantle convection is taken into consideration.

In summing up these possibilities, it becomes clear that it is time for us to discard the fixed idea of the earth as a nonflowing solid object, and reconsider the earth's thermal history from the perspective of a working hypothesis in which it is maintained that the earth flows. In this context, the prospect of combining the study of the earth's thermal history and the concepts of plate tectonics (oddly enough the two areas have thus far been pursued independently) is a most interesting one. A truly exciting synthesis can be anticipated.

As we look back on the past research conducted on the earth's history, we discover that certain basic questions, such as whether the earth was hot or cold to start with, or whether it was ever completely molten, continue to recur. Yet, one cannot brand the research that has taken place in between as pointless, for each time we have returned equipped with a deeper understanding in our search for the truth.

## The Remaining Questions

Although the prevailing view of the earth has shifted from a fixist to a mobilist one, a great many questions still remain. It is common to find that hypotheses once considered to be reasonably sound to the first order of approximation are in fact fraught with unresolved difficulties upon further examination. It may be that the theories of sea-floor spreading and plate tectonics are reaching that point. V. V. Beloussov of the USSR is one scientist who has been sharply critical of the new view. In an article entitled "Against the hypothesis of sea-floor spreading" (1970), he points out many of the difficulties inherent in the sea-floor spreading hypothesis. Although the crux of his objection is that the new view cannot explain continental geology to his satisfaction, part of his attack focuses on the proposed origin of the ocean floor. On both sides of the oceanic ridge axis are symmetrical belts of various kinds—the most typical being the geomagnetic stripes. Examination of a geomagnetic profile, such as that in Figure 3-3, will show that the anomalies are stronger near the ridge axis and weaker as one moves away from the axis. The underground structure of the mid-oceanic ridges, as estimated by seismic prospecting or by the examination of gravity in the area, likewise exhibits modulation that varies in relation to the proximity of the ridge axis: that is, the second layer is thick on the crest whereas the third layer becomes thicker as the distance from the axis increases. If the ocean floor is spreading as if on a conveyer-belt system, the present ocean floor, now hundreds of kilometers away from the ridge crest, must once have been on top of the crest. Why then, Beloussov argues, isn't the structure of the crest similar to that of the distant sea floor? This is a good point, and one that many scientists are trying to explain. There seems to be some hope of doing so by considering the changes that take place in the properties of the oceanic crust as it spreads, such as weathering, hydration, and other metamorphic processes.

Another aspect of Beloussov's attack concerns the distribution of sediment on the upper layer of the ocean floor. On the ridge crest the sediments are almost entirely absent, whereas the sediment layer thickens increasingly the further it is from the ridge. Qualitatively, this condition supports the sea-floor spreading hypothesis, since the implication is that the ridge crest is younger than the areas farther away from it. Quantitatively, however, it does not quite satisfy the requirements of the theory; specifically, the thickness of the layer does not increase constantly as the theory would lead us to suppose, but somewhat haphazardly. Beloussov cites this as another difficulty that must be resolved if the sea-floor spreading hypothesis is to be confirmed. But here again the vast information gathered by the Deep Sea Drilling Project (DSDP) has come to the rescue. The thickness of the sediments seems to decrease with increasing distance of the ocean floor from an equatorial zone of high biogenic sedimentation rate, so that the distribution of thickness deviates from a simple function of distance from ridges.

Marine sediments pose yet another question. When the ocean floor descends into the mantle at the oceanic trenches, what becomes of the sediment that is carried on top of the floor? Won't it be jammed between the oceanic and continental plates, severely folded, and then plastered on the bottom of the trenches? Observation of the trenches, however, proves otherwise. Very little sediment is to be found on the bottom of the trenches, and what little there is does not seem to have been disturbed much. If we suppose that subduction has been taking place for the past 100 million years and that the sediment thickness of the subducting ocean floor averages 200 meters, about 20 kilometers of sediment should have accumulated in the oceanic trenches by now. Moreover, the deep trench receives constant sedimentation from the land. The quantity of such trench sediments could easily be more than that carried by the oceanic plate. Altogether, there should be a great quantity of sediments in the trench. But in actuality most trenches have only a thin veneer of undeformed sediments along the bottom. What is the reason for such a discrepancy? Beloussov asks. It was an enigma for some time. But now, thanks to the powerful deep-penetrating seismic-reflection profiling techniques, the locations of the missing sediments are being disclosed. Records like that shown in Figure 2-3 are being accumulated for analysis.

Figure 6-4 shows the concept now held by scientists. The "missing" sediments are all piled up in the landward wall of the trench in a most spectacular manner. Both oceanic sediment veneer and the sediments from the land are split into thin slices by thrust faults and folded, thereby forming an intricate complex structure called the *accretionary prism* (see also Figure 5-23). It is suspected that, in many places, not only the soft sediments but also slivers of igneous basement (oceanic crust) itself are accreted to this structure.

Deformed marine sediments have been recognized by geologists for years and described by them as *geosynclinal* sediments. In many of these geosynclinal deposits have been found igneous rocks called *ophiolites*, whose origin was always a mystery. It now seems highly likely that the ophiolites are old pieces of oceanic crust thrust up against the continents along with ocean floor sediments to form the geosynclines. What actually takes place at the subduction zones will certainly be one of the most important subjects to be investigated in the years to come. The International Phase of Ocean Drilling (IPOD) is a continuation of DSDP, which terminated in 1975. In this international endeavor, drilling at subduction zones is considered to be one of the prime objectives. It is hoped that deep drilling in the walls on the continent side of the trenches will provide us with the information needed to solve these important problems.

Another question that Beloussov has labeled a mystery is that of the "migrating oceanic ridges" (see pages 102 through 109). The oceanic ridge is thought to be the place at which mantle convection ascends and the ocean floor is produced. Can such a region move about merely to satisfy the geometric requirements of plate motions? Beloussov contends that to infer further that an oceanic ridge can itself descend into an oceanic trench is completely self-contradictory. His point is well taken, but the apparent contradiction is resolved if one accepts an alternative theory about the nature of the ridges: they may not represent the upwellings of deep mantle convection currents but instead may be passive windows from which asthenospheric materials emerge to form new plates.

Even more mysterious than the migrating ridges are the so-called "fracture zones." Explaining fracture zones as transform faults was J. T. Wilson's dramatic contribution (1965). Yet Wilson himself has

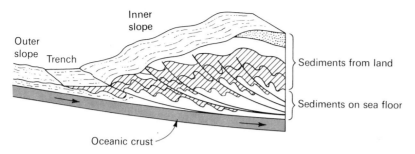

FIGURE 6-4
A model of the inner slope of a trench. The arrows indicate the movement of the ocean crust. [After D. R. Seely, P. R. Vail, and G. Walton, "A Trench Slope Model," in C. A. Burk and C. L. Drake, Eds., *The Geology of Continental Margins.* Springer-Verlag, 1974.]

not provided a thorough explanation of why the ridges were offset in the first place. If the oceanic ridge is a zone in which a convection flow upwells, why should the ridge fracture clumsily into transform faults? A smooth curve would be much more probable since the substance involved here is fluidal. Influenced by all those considerations, scientists are now inclined to believe that the ridges are indeed passive windows rather than surface manifestations of deep convection. But if ridges are not produced by convection currents in the mantle, what happens to the very basic concept that mantle convection drives the motions of the plates? To summarize, proponents of continental drift were at a loss for a driving mechanism when the idea of mantle convection as the force appeared like a Messiah to save the theory. Now, ironically enough, mantle convection seems to be on the verge of rejection by those who most strongly support the sea-floor spreading hypothesis and plate tectonics.

In a relentless barrage of criticism, Beloussov and the American petroleum geologist A. Meyerhoff have pointed out the danger of making careless generalizations in the attempt to explain a complex phenomenon on the basis of just one of its superficial manifestations. (Such explanations can be likened to that by the astronomers who observed dots on the planet Mars and linked them to demonstrate the existence of canals there.) In other words, it is all too easy to "see" with eyes already blinded by the beholder's own anticipation of what he expects to find. Beloussov's own explanation of marine geology is equally subject to criticism, however. In brief, he seems to believe that the continents are ancient and are gradually "oceanizing." The development of the ocean is caused by some activity that took place first in the peripheral regions of the present ocean. At present, "oceanization" has progressed to the region of the mid-oceanic ridge. (This explains the progressively younger age of the ocean floor towards the ridges!) Beloussov does not seem to have a convincing explanation for the physicochemical mechanism of "oceanization." He suggests that some kind of hot liquid seeps from the mantle that causes the continental crust to transform into oceanic crust. Many petrologists, however, consider such a process to be utterly impossible. And yet, to discount Beloussov's theory altogether, simply because it is theoretically difficult, would be narrow-minded, especially if a careful examination of "oceanization" from a variety of perspectives proves it to be far more plausible than other hypotheses. "Oceanization," then, is yet another working hypothesis. Who knows, it may be the prelude to a still *newer* view of the earth.

### The Driving Mechanism (1):
### A Summary of the Possibilities

Now, to return to plate tectonics, recall that identification of the driving mechanism is fundamental. What are the plausible proposals? We have seen that the simple convection current hypothesis has numerous difficulties. The problem of the horizontal scale of the cells (page 180) and that of migrating ridges (page 109) were already sufficient to make us doubt the importance of convective flow in the asthenosphere. In 1973, however, E. Artyushkov, a young Soviet geophysicist, produced an even more devastating argument against mantle convection. He argued convincingly that the viscosity of the asthenosphere, particularly under oceanic areas, should be one or two orders of magnitude smaller than the usually assumed value, estimated from the postglacial rebound (page 176) for Scandinavia and North America. Therefore, the mechanical force exerted at the bottom of the lithosphere owing to the flows in the asthenosphere would be much too weak to be significant. He maintained that, although there may indeed be flows in the asthenosphere, they are of no importance to plate motion. Whether Artyushkov was absolutely right or not, the general question of the driving mechanism seemed to require far more serious thought.

D. Forsyth and I (1975) started to seek some of the answers to this basic question in 1972 when I was a visiting professor at M.I.T. The

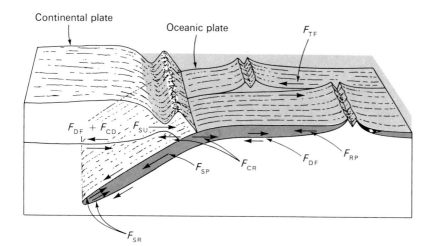

FIGURE 6-5
Possible forces acting on the lithospheric plates. [From D. Forsyth and S. Uyeda,
"On the Relative Importance of the Driving Forces of Plate Motion." *Geophys. J.*
**43**, p. 165, 1975.]

work seemed unpromising and we interrupted it until 1974, when
both of us happened to be working at the Lamont-Doherty Geological Observatory. Needless to say, the problem has not been ultimately
resolved, and a number of other scientists are working on the subject.
Although some appear to have opinions similar to ours, a good many
of them have differing views. What follows is merely a description of
our own approach to the problem.

The first step was to review the major driving mechanisms that
have been suggested in recent years by geophysicists who recognized
the difficulties inherent in simple mantle convection. These forces
are shown in Figure 6-5. Let us examine them.

**Mantle Drag Force.** It is clear that, when a large plate moves, there
must be a net mass flow in the deeper mantle from trench to ridge,
balancing the mass transport. Thus there is little doubt that flow
occurs in the mantle. Classical mantle convection theory assumes
that the flow directly below the plate drives the plate and is thus
flowing in the direction of the plate's movement, so that the return
flow must take place deeper in the mantle. But if the plate is driven by
some other force or forces, as we shall explain, the return flow can be
directly below the plate and the asthenosphere will then act as a brake

on the plate motion. Whether it is driving or resistive, the *mantle drag force* $F_{DF}$ on the plate is proportional to the area and to the velocity of the plate relative to the asthenosphere. Since the viscosity of the asthenosphere under the continents may be higher, an additional term, *continental drag* $F_{CD}$, is used for the mantle drag force that operates on the continental part of the plate.

**Ridge Push.** Although ridges may be formed passively, they have an elevated topography and deep low-density roots. Such a ridge, even though isostatically in equilibrium, has a higher potential energy than older sea floor on which there are no ridges. Thus the ridges tend to spread out, thereby producing a *ridge push force* $F_{RP}$ on both sides of the plates. Several geophysicists have contended that this force is the principal driving mechanism.

**Slab Pull and Slab Resistance.** The subducting slab underneath the trenches is considered to be colder and thus denser than the surrounding mantle. Several geophysicists have suggested that the negative buoyancy due to the density difference should pull the slab downwards, and that this *slab pull force* $F_{SP}$ is transmitted to the whole plate as a major driving force (Elsasser, 1969). Theoretical calculations indicate that the slab pull can be an order of magnitude greater than the ridge push $F_{RP}$. Since this force is due to the gravitational body force, it is simply proportional to the density difference between the slab and the surrounding mantle and is independent of the slab's velocity as long as the velocity is great enough (say, greater than 5 centimeters per year). When the downgoing velocity is smaller, heat conduction from the surrounding material will significantly lessen the temperature difference, and therefore the density difference between the downgoing slab and the surrounding mantle. As the slab plunges into the mantle, pulled by $F_{SP}$, it should meet *slab resistance* $F_{SR}$, which, for the sake of simplicity, may be designated as proportional to the velocity of underthrusting. It is thought that the resistance is concentrated at the leading edge of the slab because, at a shallow depth, the mantle is likely to be soft.

**Suction.** We have seen that, around the edge of the Pacific, trenches tend to migrate seaward, as do the overthrusting continental plates so that the area of the Pacific Ocean is decreasing while that of the Atlantic Ocean is increasing. In 1971 Elsasser explained this fact by assuming a *suction force* $F_{SU}$. Although the physical nature of this force and the degree of its importance are not altogether clear, we will list it for the sake of completeness and consider it to be independent of the velocity of plate motions.

**Colliding Resistance and Transform Fault Resistance.**   Plates are expected to experience various resistances to their motion. Slab resistance $F_{SR}$, explained above, operates at the leading edge of the slab, whereas the mantle drag forces $F_{DF}$ and $F_{CD}$ operate at the lower surface of the plate. Both these resistances occur between the plate and the deeper mantle. The resistance terms, *colliding resistance $F_{CR}$* and *transform fault resistance $F_{TF}$*, are the resistances between plates. Most shallow earthquakes are caused by these "interplate" resistances. $F_{CR}$ is the resistance at the colliding (or converging) boundary. We believe that the magnitude of these forces is independent of the relative velocity—the velocity of relative motion—between plates. It may sound strange at first to say that resistive forces are equal regardless of their relative velocity. But since earthquakes take place when the stress reaches a critical value, it may be reasonable to assume that greater relative velocity results in greater seismicity and not greater stress. The same applies to the resistive force $F_{TF}$ at the transform fault. Although the magnitude of these forces is velocity independent, the direction is controlled by the relative velocity: that is, the forces act in a direction antiparallel to that of relative plate motion. It is also worth noting that these forces acting on two plates are exactly equal in magnitude and work in the opposite direction as a consequence of Newton's law of action and reaction.

In summary, the possible driving forces are slab pull $F_{SP}$, ridge push $F_{RP}$, and suction $F_{SU}$, and the resistive forces are slab resistance $F_{SR}$, colliding resistance $F_{CR}$, and transform fault resistance $F_{TF}$. Whether the mantle drag forces $F_{DF}$ and $F_{CD}$ are driving or resistive depends on the direction of relative motions between the plate and the underlying asthenosphere.

How could we find out which of the above forces are more important than others? This could only be done through a careful examination of the actual observed motions of plates. We studied 12 plates, as shown in Figure 6-6. The direction and velocity of *relative* motions *between the plates* are determined by plate tectonic analysis as explained in Chapter 4. But since some forces ($F_{FD}$, $F_{CD}$, $F_{SP}$) are dependent on the plates' velocities relative to the *mantle*—their *absolute* velocities—we had to obtain these velocities also. To do so, we assumed that a worldwide system of *hot spots* remained fixed in space relative to the deep mantle. Now, what are hot spots?

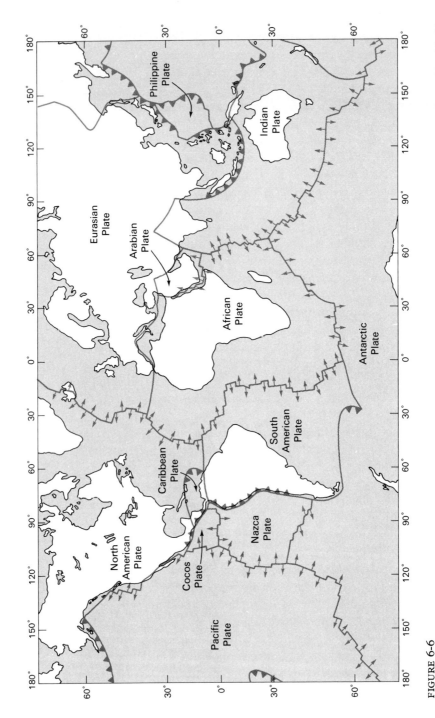

FIGURE 6-6
Twelve plates and their motions. Triangles along boundaries indicate direction of underthrusting where a downgoing slab can be identified by the occurrence of intermediate or deep focus earthquakes. Small arrows on ridge boundaries indicate approximate direction of relative motion. [After D. Forsyth and S. Uyeda, "On the Relative Importance of the Driving Forces of Plate Motion." *Geophys. J.* **43**, p. 163, 1975.]

In 1965, J. T. Wilson, the founder of the tranform fault hypothesis, made another suggestion. He noted that at certain locations on the earth, such as Hawaii and Iceland, volcanoes have been active for long periods of time. The source of magma for these volcanoes is believed to be located deep below the lithosphere so that the position of volcanic activity is fixed relative to the mantle. When a plate moves past such a magma-producing spot, the surface volcanoes are carried away with the plate, but the source continues its activity from the fixed position. As a result, a long chain of volcanoes, such as the Hawaiian volcanic chain (Figure 6-7a), will form. In fact, it was known that the age of the volcanism on the island chain increases according to their distance from the presently active island of Hawaii, situated at the southeastern end. (Figure 6-7b).

When such a spot happens to be on an actively spreading ridge, like Iceland, chains of volcanic islands or seamounts are formed on both sides of the ridge because the plates on both sides are spreading away from each other. Wilson identified several more examples of such volcanoes and their associated chains of extinct volcanoes. He called these spots *hot spots*. He proposed, in effect, that the "absolute" motions of a plate were imprinted on the sea floor in the form of a ridge made up of extinct volcanoes. Once again Wilson had presented an extremely viable, although this time theoretically unproven, hypothesis.

Later, W. J. Morgan extended this idea and demonstrated that the velocities of the "absolute" motions of plates during the Cenozoic era could be determined as shown in Figure 6-8. These are the motions that are in accord with the relative motions specified by plate tectonics and that satisfy the condition that the proposed hot spots must be stationary relative to each other and to the mantle.

Morgan, taking the idea a step further, argued that hot spots are maintained by a quite localized upwelling of the mantle convection current, and that the mantle flow associated with hot spots, upwelling in plumes and descending slowly everywhere else, may be the major driving mechanism of plate motions. Although these later speculations are subject to debate, the reasonably fixed positions of hot spots appear to be substantiated by further critical examinations.

Therefore, in our discussion on plate motions, we tentatively adopted the worldwide system of hot spots as an "absolute" frame of reference for determining the motion of the plates relative to the mantle.

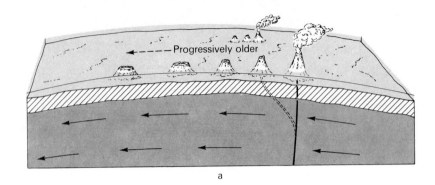

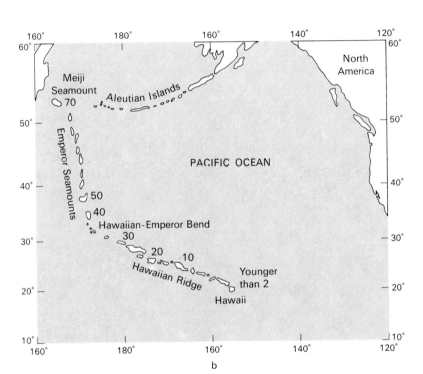

FIGURE 6-7
(a) Model demonstrating the way in which the Hawaiian volcanic island chain would have formed according to the hot-spot hypothesis. [After J. T. Wilson, "Continental Drift," Copyright © 1963 by Scientific American, Inc. All rights reserved.]
(b) Ages, in millions of years, of volcanoes in the Hawaiian-Emperor chain. [After D. A. Clague et al., "Petrology and K-Ar Ages of Dredged Volcanic Rocks from the Western Hawaiian Ridge and the Southern Emperor Seamount Chain." *GSA Bull.* **86,** p. 991, 1975.]

The absolute motions given in Figure 6-8 constituted our basic data. Our task was to deduce the driving mechanism from these data. We first noticed a remarkable regularity in the plate motions. That is, the velocity of a plate is always large when the plate has a substantial underthrusting trench boundary and vice versa. Readers will readily agree that this regularity is an outstanding one. The Cocos, Pacific, Nazca, Philippine, and Indian Plates are plates having substantial underthrusting boundaries, and for each the average velocity (the velocity averaged over the area of each plate) is between 6 and 9 centimeters per year, whereas all other plates have an average velocity of less than 4 centimeters per year, and most of those less than 2 centimeters per year. To demonstrate this regularity, we plotted the average velocity of each plate against its fractional trench length (that is, the fraction of trench length in the total boundary) as shown in Figure 6-9. This observation seemed to tell us that among the various driving forces, the trench pull $F_{SP}$ should be the most important one. We also examined the correlation between plate velocity and other geometrical factors such as the plate's area, the area of the continental part, the total length of ridges, the length of transform faults, and the length of the overthrusting side of the trench. This provided us with some more interesting clues. In particular we observed that velocity was not clearly correlated with these factors except for the continental area of a plate.

The fact that the velocity has no correlation with the area of a plate is important. If the forces acting at the plate boundaries are driving forces and the mantle drag is the major resistance, such a lack of correlation is hard to understand. For instance, the Nazca, Cocos, and Pacific Plates are very similar except in area. Therefore, if mantle drag is the principal resistive force, the Pacific Plate, having by far the greatest area, should move much more slowly. Both Morgan and McKenzie noted in the early 1970s this lack of correlation between the area and velocity of plates, and concluded that the plate velocity is determined primarily by the mantle flow, and not by forces at the boundaries.

However, we recognized an alternative interpretation: the mechanical coupling between the plate and the mantle beneath may be weak in oceanic areas, as Artyushkov maintains (page 190), so that mantle drag has little influence on the velocity of the plates. We chose this alternative as our working hypothesis.

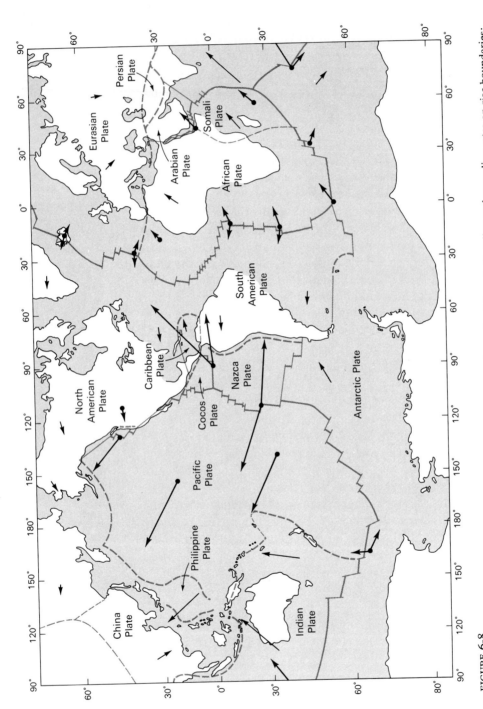

FIGURE 6-8
Present motions of plates over hot spots. The relative motions were determined from fault strikes and spreading rates on rise boundaries; with an appropriate constant rotation added, absolute motion of each plate over the mantle was determined. The lengths of arrows are proportional to the plate speed. [After J. Morgan, "Deep Mantle Convection Plumes and Plate Motions." *Amer. Assoc. Petrol. Geol. Bull.* **56**, p. 203, 1972. Redrawn with permission of the author.]

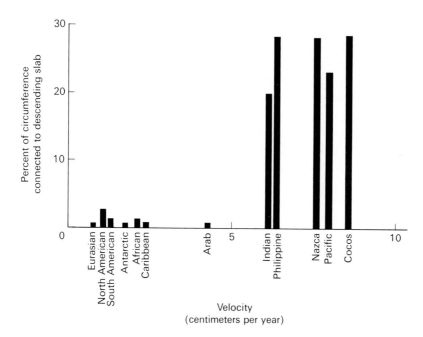

FIGURE 6-9

The percentage of plate circumference connected to the downgoing slab versus the absolute average velocity of plates. [From D. Forsyth and S. Uyeda, "On the Relative Importance of the Driving Forces of Plate Motion." *Geophys. J.* **43**, p. 163, 1975.]

We found the correlation between velocity and continental area to be significant. Plates like those carrying Eurasia, North and South America, Antarctica, and Africa all have large continental areas and their velocities are all less than 2 centimeters per year. To us this finding indicated that the resistance due to mantle drag is stronger under the continents than under the oceans.

The velocity of a plate has no obvious correlation with the total length of the ridges along its border, with the total length of transform faults, or with the total length of trench, provided the plate is on the overthrusting side of the trench. We have already seen that, for the plate in the underthrusting side of the trench, the plate velocity does correlate with trench length. This correlation indicated to us that ridges and the overthrusting sides of trenches are much less important agents in driving the plates than are the descending slabs, and that transform faults are not important resistive agents. In other words, $F_{RP}$ $F_{TF}$ and $F_{SU}$ are smaller than $F_{SP}$. This is not to say that ridges are not pushing the plates away laterally, because the plates on either side of the Mid-Atlantic Ridge are certainly moving apart, though slowly. It is only to say that probably the ridge push $F_{RP}$ is much smaller than the slab pull $F_{SP}$.

The inferences so far are that, as a driving force, the slab pull $F_{SP}$ is much greater than any other driving force, and continental drag $F_{CD}$ is the only significant resistive force. Now, if we recall Newton's basic law of motion (that forces must balance in order to keep a body in constant motion), and if we assume that the plates are essentially in constant motion, then the forces acting on each plate must balance. In our analysis we noticed that oceanic plates having a long underthrusting trench are under strong driving force $F_{SP}$, but that the resistive forces $F_{DF}$ and $F_{TF}$, are small. In order to balance the force, our conclusion was that the slab resistance $F_{SR}$ must be the main resistive force. Thus, the following model was presented. First, the body force $F_{SP}$ due to the excess mass in the downgoing slab is very large. $F_{SP}$ pulls the plate attached to it and the rate of the slab's descent into the mantle increases until this force is nearly balanced by the viscous resistive force $F_{SR}$ acting on the slab. The quite uniform rate of descent of 6 to 9 centimeters per year observed for the Pacific, Nazca, Cocos, Indian, and Philippine Plates represents the point of balance, which is, in effect, the *terminal velocity* of a dense body falling in a viscous medium. It is analogous to a man falling through the air with a parachute.

All other plates not attached to long subducting trenches are moving at a velocity of less than 4 centimeters per year. Most of these plates have large continents that are probably more strongly anchored to the deep mantle. But the fact that the Indian Plate, which has fairly large continents (India and Australia) and a long trench (the Sumatra-Java Trench) is moving fast, seems to suggest that the dominant factor in determining velocity is the presence or absence of a large descending slab and not the presence or absence of continents.

Forsyth and I have examined the validity of the above model more quantitatively and found it workable. If the model is accurate, we can say that the velocity of plates having descending slabs is determined by the balance between the two large forces $F_{SP}$ and $F_{SR}$ acting on the descending slab, and that it is almost completely independent of the surface geometry of the plate. We believe this conclusion to be an important one about the driving mechanism of plate motions. It would be an interesting project to examine the applicability of this model to the ancient plate motions deduced from plate tectonics.

Of course, the whole system is a kind of thermally convecting one. The body force $F_{SP}$ is due to the density difference, which is essentially the same type of force that drives simple convection. Thus we are not rejecting thermal convection, but saying that the plates are an important part of the convecting system and are not passively driven by an underlying convective flow system.

Our analysis did not allow us to assess the relative importance of the smaller forces $F_{RP}, F_{TF}, F_{SU},$ and $F_{DF}$ with confidence, because the magnitude of these forces is roughly comparable to the noise level of our data. Thus our next step is to tackle this problem so that we can gain a more complete understanding of the driving forces. For this purpose, *intraplate* earthquakes will provide key information, because the earthquakes occurring within a plate, though quite rare, can reveal the stress state within the plate resulting from all the forces. L. Sykes and M. Sbar at Lamont-Doherty Geological Observatory, and S. Solomon and his colleagues at M.I.T. are currently making valuable investigations in this area.

The "new view" has made a truly great contribution to our understanding of the earth, but we are still faced with innumerable important problems. Scientists of various nations have been confronting these problems dauntlessly one by one. A number of international programs have been initiated for this purpose. One is the Geodynamics Project that commenced operations in 1972. The International Phase of Ocean Drilling (IPOD), which started in 1975, is another.

In conclusion, majority opinion has just begun to shift toward a mobilist view of the earth, and a great deal remains to be studied in the future. Figure 6-10 illustrates the current version of continental drift envisioned by R. S. Dietz and J. C. Holden (1970). It shows the breakup of Pangaea and the changes in world geography throughout the ages, as viewed from the perspective of sea-floor spreading and plate tectonics. As part (f) of the figure shows, it is even possible to visualize how the world may look 50 million years from now! It would be interesting to compare this description with the drift proposed by Wegener, discussed at the very beginning of this book (see Figure 1-1). Indeed, the "insight" exhibitied by this great man remains impressive even today. Nevertheless the time may come when even the "new view" will be shown to contain some fatal flaw and so become obsolete. But at present our main challenge is to continue doing everything possible to pursue its implications for all branches of earth science, and to test them in the crucibles of experiment and observation. If we are successful, we may yet arrive at a comprehensive understanding of the earth.

FIGURE 6-10

(a) The ancient land mass Pangaea as it may have looked 200 million years ago. Panthalassa, the ocean surrounding Pangaea, evolved into the present Pacific Ocean, and the present Mediterranean Sea is a remnant of the Tethys Sea.

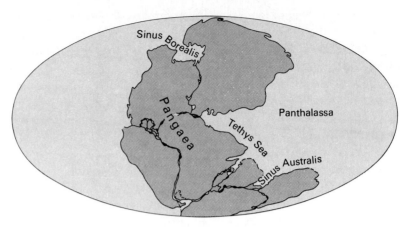

a  200 million years ago

(b) World geography at the end of the Triassic period, 180 million years ago, after about 20 million years of drift. The land mass has now become two supercontinents Laurasia and Gondwana. The light gray areas represent the new ocean floor. Spreading zones are represented by heavy lines, transform faults by fine lines, and subduction zones by hatched lines (where a line is broken, this indicates some uncertainty that the feature was present at the time). Arrows depict motions of continents since drift began.

(c) World geography at the end of the Jurassic period, 135 million years ago, after about 65 million years of drift.

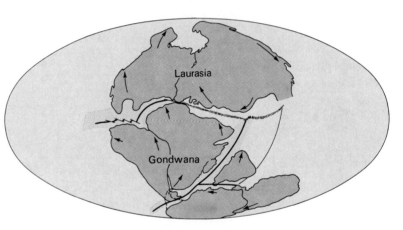

b  180 million years ago

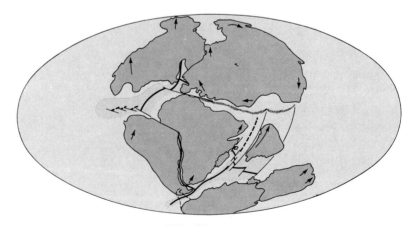

c  135 million years ago

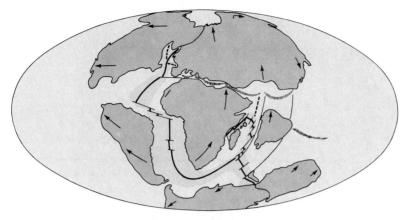

(d) World geography at the end of the Cretaceous period, 65 million years ago, after some 135 million years of drift.

d  65 million years ago

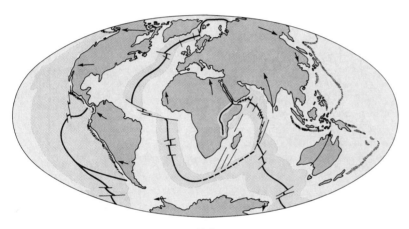

(e) World geography today, showing sea floor produced during the past 65 million years, in the Cenozoic period.

e  Today

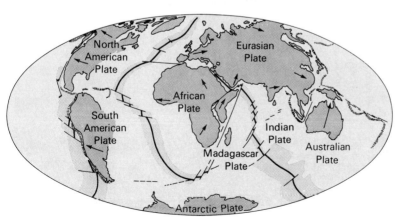

(f) World geography as it may look some 50 million years from now if present-day plate movements continue. [Parts (a) through (f) after R. S. Dietz and J. C. Holden, "The Breakup of Pangaea." Copyright © 1970 by Scientific American, Inc. All rights reserved.]

f  50 million years from today

# Typical rocks of the continental crust, oceanic crust, and upper mantle

| Rock | Location | Density (grams per cubic centimer) |
|---|---|---|
| Granite | Upper continental crust | 2.7 |
| Basalt | Oceanic crust, probably lower continental crust | 2.9 |
| Eclogite (heavy rock, high-pressure form of basalt) | Possibly upper mantle | 3.4 |
| Peridotite (heavy green-ish rock) | Probably upper mantle | 3.2 |
| Serpentinite (hydrated form of peridotite) | Possibly lower crust | 2.6 |

| Seismic P-wave velocity (kilometers per second) | Mineral content (volume percent) | Composition (weight percent) |
|---|---|---|
| 5.8–6.2 | 30% orthoclase<br>30% quartz<br>25% plagioclase<br>15% biotite<br>15% hornblende<br>15% others | 70% silicon dioxide<br>15% aluminum oxide<br>4% potassium oxide<br>4% sodium oxide<br>2% calcium oxide<br>5% others |
| 6.4–7.0 | 50% plagioclase<br>35% pyroxene<br>5% olivine<br>10% others<br>(iron oxide, etc.) | 48% slilicon dioxide<br>18% aluminum oxide<br>10% calcium oxide<br>8% magnesium oxide<br>6% ferrous oxide<br>3% ferric oxide<br>3% sodium oxide<br>4% others |
| 8 | 45% garnet<br>45% pyroxene<br>10% others<br>(amphibole, etc.) | Same as basalt |
| 8 | 85% olivine<br>10% pyroxene<br>5% others<br>(spinel, garnet, etc.) | 44% silicon dioxide<br>37% magnesium oxide<br>6% ferrous oxide<br>5% aluminum oxide<br>4% calcium oxide<br>2% ferric oxide<br>2% others |
| 6–7 | Mainly serpentine | Same as peridotite, with about 10% water |

# Bibliography

**Note:** Works designated by an asterisk (*) are especially suited for lay readers; those designated by two asterisks are collections of relevant articles.

Adams, L. H., "Some Unsolved Problems of Geophysics." *Transactions of the American Geophysical Union,* v. 28, no. 5, 1947, pp. 673–679.

Artyushkov, E. V., "Stresses in the Lithosphere Caused by Crustal Thickness Inhomogeneities." *Journal of Geophysical Research,* v. 78, 1973, pp. 7657–7708.

Atwater, T., "Implications of Plate Tectonics for the Cenozoic Tectonic Evolution of Western North America." *Bulletin of the Geological Society of America,* v. 81, 1970, pp. 3513–3536.

Beloussov, V. V., "Against the Hypothesis of Sea-Floor Spreading." *Tectonophysics,* v. 9, 1970, pp. 489–511.

Birch, F., "Elasticity and Constitution of the Earth's Interior." *Journal of Geophysical Research,* v. 57, 1952, pp. 227–286.

Blackett, P. M. S., "A Negative Experiment Relating to Magnetism and the Earth's Rotation." *Philosophical Transactions,* Royal Society of London, A245, 1952, pp. 309–370.

**Blackett, P. M. S., E. Bullard and S. K. Runcorn, Eds., *A Symposium on Continental Drift.* London: The Royal Society, 1965, 323 pp.

Bullard, E. C., J. E. Everett and A. G. Smith, "The Fit of the Continents around the Atlantic." In P. M. S. Blackett, E. Bullard and S. K. Runcorn, Eds., *A Symposium on Continental Drift. Philosophical Transactions,* Royal Society of London, A258, 1965, pp. 41–51.

**Cox, A., Ed., *Plate Tectonics and Geomagnetic Reversals.* San Francisco, W. H. Freeman and Company, 1973, 702 pp.

*Cox, A., G. B. Dalrymple, and R. R. Doell, "Reversals of the Earth's Magnetic Field." *Scientific American,* February, 1967, pp. 44–54.

Dewey, J. F. "Continental Margins: A Model for Conversion of Atlantic Type to Andean Type." *Earth and Planetary Science Letters,* v. 6, 1969, pp. 189–197.

*Dewey, J. F., "Plate Tectonics." *Scientific American,* May, 1972, pp. 56–58.

Dewey, J. F., and J. M. Bird, "Mountain Belts and the New Global Tectonics." *Journal of Geophysical Research,* v. 75, 1970, pp. 2625–2647.

Dietz, R. S., "Continent and Ocean Basin Evolution by Spreading of the Sea Floor." *Nature,* v. 190, 1961, pp. 854–857.

*Dietz, R. S., and J. C. Holden, "The Breakup of Pangaea."*Scientific American,* v. 223, no. 4, 1970, pp. 30–41.

Elsasser, W. M. "Convection and Stress Propagation in the Upper Mantle." S. K. Runcorn, Ed., *The Application of Modern Physics to the Earth and Planetary Interiors.* New York, Wiley Interscience, 1969, pp. 223–246.

Forsyth, D., and S. Uyeda, "On the Relative Importance of the Driving Forces of Plate Motion." *Geophysical Journal,* Royal Astronomical Society, v. 43, no. 1, 1975, pp. 163–200.

Gilbert, W., *De Magnete.* London; Short, 1600; New York, Dover Publications, 1958, 368 pp.

Grow, J. A., and T. Atwater, "Mid-Tertiary Tectonic Transition in the Aleutian Arc." *Bulletin of the Geological Society of America,* v. 81, 1970, pp. 3715–3722.

*Heirtzler, J. R., "Sea-floor Spreading." *Scientific American,* June, 1968, pp. 60–70

Heirtzler, J. R., X. Le Pichon, and J. G. Baron, "Magnetic Anomalies over the Reykjanes Ridge." *Deep Sea Research,* v. 13, 1966, pp. 427–443.

Heirtzler, J. R., G. O. Dickson, T. M. Herron, W. C. Pitman, III, and X. Le Pichon, "Marine Magnetic Anomalies, Geomagnetic Field Reversals and Motions of the Ocean Floor and Continents." *Journal of Geophysical Research,* v. 73, 1968, pp. 2119–2136.

Hess, H. H., "History of Ocean Basins." In A. E. J. Engel, H. L. James, and B. F. Leonard, Eds., *Petrologic Studies: A Volume in Honor of A. F. Buddington,* Boulder: Geological Society of America, 1962, pp. 599–620.

Holmes, A., *Principles of Physical Geology.* London: T. Nelson and Sons, 1945, New York: Ronald Press, 2nd ed., 1965.

*Hurley, P. M., "The Confirmation of Continental Drift." *Scientific American,* April, 1968, pp. 52–64.

Isacks, B., J. Oliver, and L. Sykes, "Seismology and the New Global Tectonics." *Journal of Geophysical Research,* v. 73, 1968, pp. 5855–5899.

Jeffreys, H., *The Earth.* Cambridge: Cambridge University Press, 5th ed., 1970, 525 pp.

Karig, D. E., "Evolution of Arc Systems in the Western Pacific." *Annual Review of Earth and Planetary Sciences,* v. 2, 1974, pp. 51–75.

Larson, R. L. and C. G. Chase, "Late Mesozoic Evolution of the Western Pacific Oceans." *Bulletin of the Geological Society of America,* v. 83, 1972, pp. 3627–3644.

Larson, R. L., and W. C. Pitman, III, "Worldwide Correlation of Mesozoic Magnetic Anomalies, and Its Implications." *Bulletin of the Geological Society of America,* v. 83, 1972, pp. 3645–3661.

Le Pichon, X., "Sea-Floor Spreading and Continental Drift." *Journal of Geophysical Research,* v. 73, 1968, pp. 3661–3697.

Le Pichon, X., J. Francheteau, and J. Bonnin, *Plate Tectonics.* Amsterdam, Elsevier, 1973, 300 pp.

McElhinny, N. W., *Paleomagnetism and Plate Tectonics.* London: Cambridge University Press, 1973, 357 pp.

McKenzie, D. P., and R. L. Parker, "The North Pacific: An Example of Tectonics on a Sphere." *Nature,* v. 216, 1967, 1276–1280.

McKenzie, D. P., and W. J. Morgan, "The Evolution of Triple Junctions." *Nature,* v. 224, 1969, pp. 125–133.

**Maxwell, A. E., Ed., *The Sea: Ideas and Observations on Progress in the Study of the Seas.* New York: Wiley-Interscience, v. 4, part 1, 1971, 664 pp.

Morgan, W. J. "Rises, Trenches, Great Faults, and Crustal Blocks," *Journal of Geophysical Research,* v. 73, 1968, pp. 1959–1982.

Morgan, W. J., "Deep Mantle Convection Plumes and Plate Motions." *American Association of Petroleum Geologists Bulletin,* v. 56, no. 1, 1972, pp. 203–213.

Néel, L., "L'Inversion d'Aimantation Permanente des Roches." *Annales de Géophysique,* v. 7, 1951, pp. 90–102.

**Phinney, R. A., Ed., *The History of the Earth's Crust.* Princeton: Princeton University Press, 1968, 244 pp.

Press, F., and R. Siever, *Earth.* San Francisco: W. H. Freeman and Company, 1974, 945 pp.

**Runcorn, S. K., Ed., *Continental Drift:* New York: Academic Press, 1962, 338 pp.

Sclater, J. G., and J. Francheteau, "The Implications of Terrestrial Heat Flow Observations on Current Tectonic and Geochemical Models of the Crust and Upper Mantle of the Earth." *Geophysical Journal,* Royal Astronomical Society, v. 20, 1970, pp. 509–542.

Sugimura, A., and S. Uyeda, *Island Arcs: Japan and Its Environs.* Amsterdam: Elsevier, 1973, 246 pp.

Takeuchi, H., S. Uyeda, and H. Kanamori, *Debate About the Earth.* San Francisco: Freeman, Cooper, 1970, 281 pp.

Vine, F. J., "Spreading of the Ocean Floor; New Evidence." *Science,* v. 154, 1966, pp. 1405–1415.

Vine, F. J. and D. H. Matthews, "Magnetic Anomalies over Oceanic Ridges." *Nature,* v. 199, 1963, pp. 947–949.

Vine, F. J., and J. T. Wilson, "Magnetic Anomalies over a Young Oceanic Ridge off Vancouver Island." *Science,* v. 150, 1965, pp. 485–489.

★Wegener, A., *The Origin of Continents and Oceans.* London: Methuen, 1924; New York: Dover, paperback ed., 1966.

Wilson, J. T., "A New Class of Faults and Their Bearing on Continental Drift." *Nature,* v. 207, 1965, pp. 343–347.

★★Wilson, J. T., Ed., *Readings from Scientific American: Continents Adrift and Continents Aground.* San Francisco: W. H. Freeman and Company, 1976, 230 pp.

Wyllie, P. J., *The Dynamic Earth: Textbook in Geosciences.* New York: Wiley, 1971, 416 pp.

# Index

ABE, K., 157
Absolute age of rocks, 10, 39
Absolute velocities, 193
Accretionary prism, 188
ADAMS, L. H., 1, 19, 26, 206
Air-gun method, 48
AKI, K., 1
American Miscellaneous Society, 59
ANDERSON, D., 94, 184
Anticlines, 20
Appalachians, 122
ARONSON, J., 184
ARTYUSHKOV, E. V., 190, 197, 206
Asthenosphere, 65, 94, 111, 180, 190
ATWATER, T., 107, 108, 118, 206

BABA, K., 151
Back-arc basins, 129
BACKUS, G., 27
BACON, F., 6
BARAZANGI, M., 54, 134
BARON, J. G., 207
Basalt, 15, 52, 90, 149
Basin and Range Province, 107
BECK, R. H., 49
BECQUEREL, A., 182
BELOUSSOV, V. V., 67, 167, 186–190, 206
BÉNARD, H., 174–176
Bénard cells, 174

BENIOFF, H., 133
Benioff zone, 133
BIRCH, F., 28, 184, 206
BIRD, J., 121, 163, 207
BLACKETT, P. M. S., 26, 32, 206
BLOCK, M. J., 175
BRUNHES, B., 30
BULLARD, SIR EDWARD, 27, 39, 50, 52, 206
BYERLY, P., 78

Caledonian Mountains, 122
CAREY, S. W., 18, 125
Cenozoic sequence, 116
CHAMELAUN, F., 30
CHANDRASEKHAR, S., 175, 178
CHAPMAN, S., 156
CHASE, C. G., 114, 116, 208
Chronology of geomagnetic field reversals. See Earth's magnetic field
Circum-Pacific orogenic belt. See Orogenic belt
CLAGUE, D. A., 196
CLARK, S. P., 183
Colliding resistance, $F_{CR}$, 192
Continental arcs, 57
Continental drag, $F_{CD}$, 192
Continental drift, 6, 19, 201
Continental drift theory, 6
    death of, 23
    revival of, 36

Continental margins, 122
  active, 124
  Atlantic-type, 122
  Pacific-type, 122, 125
  passive, 124
Contraction theory, 19
Convection
  aspect ratio, 180
  Bénard-Rayleigh, 175
  in the mantle, 37, 127, 173, 176,
    184, 188, 191
  thermal, 37, 173, 201
COOPER, A., 168
Core, 24, 28, 178, 184
Corer, 73
COX, A., 30, 69, 70, 71, 206
Crust, 13, 24, 64
  continental, 15
  Japan Trench, 147
  oceanic, 16, 52, 64
  Pacific Basin, 147
  Sea of Japan, 146
CURIE, M., 182
Curie point, 26, 29

DALRYMPLE, B., 30, 69, 206
Darwin Rise, 92
Declination. See Earth's
  magnetic field
Deep Sea Drilling Project
  (DSDP), 16, 81, 84–87, 92,
  116, 187, 188
DEWEY, J., 96, 121, 123, 163, 207
DICKINSON, W., 150
DICKSON, G. O., 207
DIETZ, R., 64, 173, 201, 203, 207
Dipole field, 32
DOELL, R., 30, 69, 206
Driving mechanism
  of continental drift, 22, 38
  of plate tectonics, 190, 197
DU TOIT, A. L., 18
Dynamo theory, 27

Earth
  cold-origin hypothesis of, 184
  electrical conductivity of, 183
  expanding, 125
  hot-origin hypothesis of, 20, 23,
    183
  structure of, 24
  thermal history of, 182
Earthquake epicenters
  Japanese area, 136
  world distribution of, 53, 54,
    82–83, 93, 134
Earthquake magnitude, $M$, 142
Earthquakes, 93, 132
  deep-focus, 1, 105, 122, 133,
    144, 149, 157
  great, 143
  intermediate-focus, 105, 133
  interplate, 144
  intraplate, 144, 201
  shallow-focus, 142, 193
Earthquake source mechanism,
  78, 97, 135
Earth's interior
  electrical conductivity of, 27,
    156
  temperature in, 1
  thermal conductivity of, 50, 184
  viscosity of, 190
Earth's magnetic field, 24, 32
  declination of, 29
  externally induced, 154
  history of, 28, 32
  origin of, 1, 25
  reversal history of, 30, 69, 81,
    117
  reversal of, 30, 69
  of sea floor, 58
  secular variation of, 29
  time variation of, 154
East Pacific Rise, 53, 70, 77, 103,
  107
Echo-sounding method, 45
Eclogite, 16, 64
Electrical conductivity, 154
  in the earth's interior, 27, 156
  under the island arc of Japan,
    156
ELSASSER, W. M., 17, 192, 207
Eltanin, 72
Eltanin-19 profile, 71, 72
Emperor Seamounts, 102, 114,
  196
ENGEL, A. E. J., 207
Epoch. See Magnetic polarity
  epoch
Erosion, 12
Events, 70
  Olduvai event, 70
EVERETT, J. E., 206
Evolution, 10
EWING, M., 44, 53
Explosion seismology, 145
  marine, 48
Extensional opening of marginal
  basins, 169

FEDOTOV, S. M., 143
Ferromagnetism, 26
Folding, 20
FORSYTH, D., 190, 191, 194, 199,
  200, 207
Fossa Magna, 159, 161
Fossil faults, 100
Fossil magnetization, 29
Fossils, 9
Fracture zones, 59, 75, 79, 188
FRANCHETEAU, J., 112, 209

Free-air correction. *See* Gravity
  anomaly
French American Mid-Ocean
  Undersea Study (FAMOUS),
  113
Front of the volcanic belts,
  147–149, 152
Frontal arc, 169
FUJII, N., 163
FUKAO, Y., 184

Gabbro, 15, 52, 64
Gammas, 67
Geodynamics Project, 201
Geomagnetic anomalies, 58
  in the East Pacific, 58, 90
  around Japan, 90
  local, 58
  in the northwestern Pacific, 91
  in the ocean, 58, 80, 82–83
  regional, 58
  of the Sea of Japan, 170
  off Vancouver Island, 60, 77
Geomagnetic field. *See* Earth's
  magnetic field
Geomagnetism. *See* Earth's
  magnetic field
Geophysical Laboratory of
  Carnegie Institute, 149
Geopoetry, 62
Geosynclinal sediments, 188
Geosyncline, 1, 20, 22, 188
Geothermal gradient, 50
GILBERT, W., 24, 25, 207
GIRDLER, R., 106
Glaciation, 17
GLASS, B., 72
*Glomar Challenger,* the, 85
Gondwanaland, 18, 22, 41
GRAHAM, J., 31
Granite, 15, 24, 52
Gravitational energy
  for accretion, 185
  for core formation, 184
Gravity
  at sea, 42, 58
  surface ship gravity meter, 48
Gravity anomaly, 43, 129–132
  free-air, 130
  seaward of trenches, 132
  in trench areas, 131
Great Magnetic Bight, 102, 114
GRIGGS, D., 127, 173, 179
GROW, J. A., 118, 207
Gulf of Alaska, 105, 114
GUTENBERG, B., 142
Guyots, 92

*Hakuho-Maru,* the, 44
HALLAM, O., 120, 121

HAMILTON, W., 107
Haruna, Mt., 30
HASEBE, K., 163
Hawaiian archipelago, 67, 147
Hawaiian volcanic chain, 195,
  196
HAYES, D., 103
HAYS, J., 121
Heat flow. *See* Terrestrial heat
  flow
Heat flow unit, 164
HEEZEN, B., 45, 53, 55, 56, 125
HEIRTZLER, J., 72, 79, 80, 81, 207
HERRON, T. M., 207
HERZENBERG, A., 27
HESS, H., 62, 92, 173, 207
Himalayas, 21
HOLDEN, J. C., 201, 207
HOLMES, A., 19, 37, 38, 62, 68,
  127, 173, 207
HONDA, H., 78, 135, 139
Honshu, 128
HORAI, K., 151, 161
Hot-origin hypothesis. *See* Earth
Hot spot, 66, 193, 195
HURLEY, P., 39, 40, 207
Hydrothermal activity, 55
Hydrothermal system, 112

Ice Age, 17, 176
Iceland, 195
Igneous activity, 1, 124
Indian Ocean floor, 56
Indonesian archipelago, 42, 58
Interarc basins. *See* Marginal seas
International Geophysical Year,
  56
International Indian Ocean
  Expedition, 56
International Phase of Ocean
  Drilling (IPOD), 90, 188,
  201
Interquake period, 143
IRVING, E., 35
ISACKS, B., 140, 141, 159, 208
ISEZAKI, N., 170, 171
Island arcs, 125
  Aleutian, 118, 168
  Circum-Pacific, 132
  drifting of, 168
  East Japan, 128, 147
  Honshu, 127, 166
  Indonesia, 58
  Izu-Bonin-Marianas, 127, 152,
    159
  Japanese, 127–128
  Kurile, 127, 152
  New Hebrides, 169
  Northeast Honshu, 152, 156,
    157
  Ryukyu, 127, 152

Island arcs (*continued*)
  Southwest Honshu, 152, 157
  thermal state beneath, 151
  Tonga, 159
  West Japan, 128, 147
Isochrons, 81
  Cenozoic, 114
  Mesozoic, 116
Isostasy, 17, 132, 176

JAMES, H. L., 207
Japan Basin. *See* Sea of Japan
Japan Sea. *See* Sea of Japan
Japan Trench, 127, 131
  crust, 147
Japanese islands
  electrical conductivity under,
    156
  underground structure of, 145
  upper mantle underneath, 157
JEFFREYS, SIR HAROLD, 23, 208
JOIDES (Joint Oceanographic
    Institutions Deep Earth
    Sampling), 85

KANAMORI, H., 143, 157, 209
KARIG, D., 168, 169, 170, 208
KATSUMATA, M., 158
KAWAI, N., 166
Keathly set. *See* Magnetic
  lineations
KELVIN, LORD, 182
KOIZUMI, K., 86
KÖNIGSBERGER, J. G., 29
KONO, Y., 151
KÖPPEN, W., 9, 35
KREICHGAUER, D., 35
KUNO, H., 149, 150
Kurile Trench. *See* Oceanic
  trenches
KUSHIRO, I., 149

Lamont-Doherty Geological
  Observatory, (Lamont
  Geological Observatory), 44,
  84, 151
Land-bridge theory, 12, 17
LARSON, R., 114, 115, 116, 117,
  118, 120, 208
Law of faunal assemblage, 9–10
Law of superposition, 9
LE PICHON, X., 95, 101, 207, 208
LEONARD, B. F., 207
LISTER, C., 112
Lithosphere, 65, 95, 110, 172
LOWES, F. J., 27
LUBIMOVA, E. A., 183

MACDONALD, G. J. F., 183
Magma, 149–151
  alkaline basalt, 149

basaltic, 110
  depth of production, 149
  parental, 148, 150
  primary, 149
  tholeiite basalt, 149
Magnetic anomalies. *See*
  Geomagnetic anomalies
Magnetic lineations, 59, 67, 69,
  81, 98, 99, 186
  ages of, 87
  in Aleutian Basin, 168
  Atlantic, 115
  Cenozoic, 114, 115
  chronology of, 81, 117
  in the East Pacific, 90, 100,
    102, 109
  in Gulf of Alaska, 114
  Hawaiian, 114
  Japanese, 91, 114
  Keathly set of, 116
  Mesozoic, 114, 116, 168
  off North America, 107
  in the northeastern Pacific, 77,
    104
  Phoenix, 114
  in Sea of Japan, 170
  in Shikoku Basin, 170
  south of Aleutians, 91
  in the western Pacific, 91
Magnetic polarity epoch, 69–71
  Brunhes normal, 70
  Matuyama reversed, 70
Magnetic poles, 25
Magnetic quiet zones, 84, 114
  Cretaceous, 116, 120
  Jurassic, 116
Magnetic stripes. *See* Magnetic
  lineations
Magnetization
  fossilized, 29
  natural remanent, 32
  remanent, 32, 72
  reverse thermoremanent, 30
  thermoremanent, 29, 110
Magnetometer, 26
  nuclear resonance type, 48
  ocean-bottom, 51
  proton-precession, 48
Magnitude. *See* Earthquake
  magnitude
Mantle, 15, 24, 37, 64, 177, 178
  convection. *See* Convection in
    the mantle
  drag force, $F_{DF}$, 191–193, 197
  material, 24
  partial melting of, 94, 110, 157
  seismic wave velocity in, 94
Marginal basins
  active, 169
  extensional opening of, 169
  inactive, 169

Marginal seas, 129, 152, 166, 171
  Bering Sea (East and West),
    154, 168
  China Sea (East and West),
    154, 166
  Fiji Basins (North and
    South), 154
  Fiji Plateau, 154
  interarc, 168
  Marianas Trough, 169
  Okinawa Trough, 152
  origin of, 166
  Parece Vela Basin, 169
  Philippine Sea (East and West),
    129, 154, 157, 166, 169
  Sea of Japan, 129, 152, 166, 170
  Sea of Okhotsk, 129, 152, 166
  Shikoku Basin, 169, 170
MARLOW, M., 168
MASON, R., 59
MATSUDA, T., 161
MATSUI, T., 185
MATTHEWS, D. H., 68, 209
MATUYAMA, M., 30, 43, 70
MAXWELL, A. E., 87, 208
McCONNELL, R., 184
McDOUGALL, I., 30
McELHINNY, N. W., 34, 36, 208
McKENZIE, D. P., 95, 97, 99, 105,
    107, 135, 197, 208
Median Tectonic Line, 159
MEINESZ, V., 42, 48, 58, 127, 129
MENARD, H. W., 92, 100
Mesozoic sequence
    (M-sequence), 116, 168
Metamorphic belts
  Hida, 161
  high-pressure low-temperature,
    160
  low-pressure high-temperature,
    160
  pair of, 161
  Ryoke, 160
  Sanbagawa, 160
  Sangun, 161
Metamorphism, 122, 124, 161
Meteor, the, 51
MEYERHOFF, A., 189
Mid-Atlantic Ridge, 51–53, 55,
    65, 87, 113, 199
Mid-oceanic ridges. See Oceanic
  ridges
MIYASHIRO, A., 112, 160, 161
MIZUTANI, H., 151, 184, 185
MOGI, K., 143, 144
Mohorovičić discontinuity (Moho
    or M discontinuity), 15, 24,
    64, 94
MOLNAR, P., 140
MORGAN, W. J., 95, 99, 105, 107,
    195, 197, 198, 208

MORLEY, L., 68
Mountain building, 120
Mountain ranges (chains), 1, 12,
    22, 97
MURAUCHI, S., 146, 168

NAGATA, T., 29, 30
NAKAMURA, K., 168
NAKANO, H., 78
NÉEL, L., 29, 31, 208
New global tectonics, 172
Newtonian fluid, 177, 179

Ocean bottom seismometer, 51
Ocean floor, 63
  age of, 81, 87–89
Ocean-floor spreading. See
    Sea-floor spreading
Oceanic ridges, 51, 62, 82–83, 95,
    105, 109, 132, 186
  Juan de Fuca and Gorda, 77,
    107, 109
  Kula-Pacific, 118
  Mid-Atlantic, 51, 65, 87, 113,
    199
  migration of, 102, 188
Oceanic trenches, 51, 56, 63, 97,
    192
  Aleutian, 57, 91, 102
  Izu-Bonin-Marianas, 57, 127
  Japan, 127
  Java-Sumatra, 200
  Kurile, 57, 90
  negative gravity anomaly of,
    131
Oceanization, 167, 190
Okinawa Trough. See Marginal
  seas
OLIVER, J., 159, 208
OPDYKE, N., 72, 73
Ophiolites, 188
Orogenesis, 19, 121, 161, 163,
    173
Orogenic belt (zones), 122, 159,
    161
  Alpine-Himalayan, 132
  Circum-Pacific, 105
  Paleozoic, 121
Orogenic cycles, 122
Orogeny, Pacific-type, 163, 165

Paleomagnetic stratigraphy, 73
Paleomagnetism, 28, 32
  of sea floor, 116
  of seamounts, 119
Pangaea, 6, 122
  breakup of, 201, 202
PARKER, R. L., 95, 97, 208
Partial melting, 94, 110, 157
PEARSON, J. R. A., 175
Peridotite, 16, 24, 52, 63

PETER, G., 102
Petrographic provinces, 149
Philippine Sea. *See* Marginal seas
PHINNEY, R. A., 208
Pillow lava, 110
PITMAN, W. C. J., 72, 79, 83, 88, 103, 104, 118, 120, 207, 208
Plate, 95
    Antarctic, 100
    Cocos, 197
    Farallon, 108, 118
    Indian, 197
    Kula, 118, 168
    Nazca, 197
    North American, 97
    Pacific, 97, 100, 116, 197
    Philippine, 197
Plate boundary, 95
    accreting, 97
    consuming, 97
    converging, 97
    diverging, 97
    transform, 97
Plate tectonics, 93
    driving mechanism of, 190, 197
Plumes, 195
Polar wandering, 34–36
Polarity epoch, 69
Pole
    earth's rotational, 118
    geomagnetic, 32
    paleoclimatological, 35
    paleomagnetic, 35
    of rotation between plates, 97, 98, 101
Pole-fleeing force, 23
Post-glacial rebound, 176, 190
Potassium-argon method, 69
Precambrian Era, 10
Precision depth recorder, 44
PRESS, F., 11, 33, 94, 95, 208
Proto-Atlantic, 121

Radiative conductivity. *See* Thermal conductivity
Radioactive elements, 10
    radioactive potassium, $K^{40}$, 69
Radioactivity, 182
RAFF, A., 59
RAYLEIGH, LORD, 174
Rayleigh number, 175, 177
    critical, 175
Red Sea, 56
Regression, 120
Relative motion between plates, 99, 193
Remanent magnetism
    natural, 32
    of sediments, 72
Remnant arc, 169

Research Group for Explosion Seismology, 145
REVELLE, R., 44
RICHTER, C., 142
Richter magnitude scale, 142
Ridge push, $F_{RP}$, 192
RIKITAKE, T., 27, 31, 156
RINGWOOD, A. E., 150
Rocks
    extrusive, 16
    igneous, 10
    intrusive, 16
    metamorphic, 112, 159
    sedimentary, 11
    volcanic, 29, 148
RUNCORN, S. K., 32, 35, 178, 206

SAKATA, M., 180
San Andreas Fault, 76, 77, 97, 100, 106
Satellite navigation, 51
SBAR, M., 201
Scandinavian peninsula, 176
SCHOLL, D., 168
SCLATER, J., 111, 112, 113, 209
Scripps Institution of Oceanography, 44, 84, 151
Sea-floor spreading, 64
    diffuse, 171
    history of, 116
    pulse of, 120
    rate of, 70, 81, 99, 120
Sea of Japan, 129
    crust, 146
    heat flow of, 152
    Japan Basin, 146
    magnetic survey of, 171
    origin of, 165
    upper mantle of, 157
Sea of Okhotsk. *See* Marginal seas
Second layer, 16, 90
SEELY, D. R., 189
Seismic energy, 142
Seismic plane. *See* Benioff zone
Seismic prospecting, 145
Seismic reflection, 48
Seismic waves, 13, 24, 157
    *P* waves, 13–15, 78, 138
    *S* waves, 13–14, 94, 138
    surface waves, 157
    velocity of, 16, 52
Self-reversal of rock magnetism, 30
Serpentinite, 16, 52, 64
Serpentinized peridotite, 63
SHIDA, J., 78
Shikoku Basin, 169
SHIMAZU, Y., 27
Short-lived radio isotopes, 185

SHOUTEN, H., 116
SIEVER, R., 208
Sinking slab model, 197
Slab, 135, 142, 158–159, 163
    pull, $F_{SP}$, 192
    resistance, $F_{SR}$, 192
SMITH, A. G., 206
SNIDER, A., 6
Solidus temperature, 94, 111
SOLOMON, S., 201
*Spencer F. Baird,* the, 53
Subduction, 97, 122, 126, 135,
    144, 147, 159, 173, 187
    speed of, 99
    zones of, 188
Suction force, $F_{SU}$, 192
SUGIMURA, A., 127, 128, 129, 148,
    161, 165, 209
Surface-ship gravity meters. *See*
    Gravity
Surface tension, 175
SYKES, L., 79, 143, 201, 208
Synclines, 20

TAKEUCHI, H., 27, 161, 162, 180,
    181, 209
TALWANI, M., 44, 132
TARLING, D., 30
TERADA, T., 168
Terminal velocity, 200
Terrestrial heat flow, 50, 151
    in Japanese area, 151, 153
    in Korea, 151
    on land, 50
    in marginal seas, 152, 154
    on mid-oceanic ridges, 53
    in north Pacific, 111
    at sea, 52, 53, 55, 111
    in South America, 153
    in trench areas, 58, 154
    in western Pacific area, 154,
    155
    world average of, 151
THARP, M., 45, 56
THELLIER, E., 29
Thermal conductivity, 50
    radiative, 183, 184
Thermoremanent magnetism.
    *See* Magnetization
Third arc, 169
Third layer, 16, 90
TOMITA, T., 149
TOMODA, Y., 48, 130, 170
Topography of the Pacific sea
    floor, 112
TOZER, D., 184, 185

Transcurrent fault, 74–75
Transform fault, 74–75, 82–83,
    99, 188
    direction of, 100
    resistance, $F_{TF}$, 192
    verification of, 77
Transgression, 120
Trenches. *See* Oceanic trenches
Triple junctions, 105, 107
TSUBOI C., 48

United States–Japan Science
    Cooperation Program, 151
Upper mantle, 16, 63, 151
    electrical conductivity, 154
    temperature distribution, 157
    thermal state, 154, 156
Upper Mantle Project, 56
UTSU, T., 158
UYEDA, S., 91, 207, 209

VACQUIER, V., 59, 77, 119
VINE, F., 60, 68, 69, 71, 72, 77,
    102, 107, 209
Vine-Matthews-(Morley)
    hypothesis, 69
Viscosity, 174, 177, 181, 190
VOGT, P., 116
Volcanic front, 147–149, 152
Volcanoes, 1, 147
    distribution of, 147, 148
VON HERZEN, R. P., 53, 55

WADATI, K., 133
Wadati-Benioff Zone, 136, 149,
    158
WATANABE, T., 151, 153, 155
WATTS, A., 131, 132, 170
WEGENER, A., 6–9, 12–13, 21–23,
    35, 201, 209
WEISSEL, J., 170
WILKINSON, I., 27
WILSON, J. T., 65, 66, 67, 69, 74,
    77, 121, 173, 188, 195, 196,
    209
Working hypothesis, 182
World Wide Standard
    Seismograph Network
    (WWSSN), 78
WYLLIE P. J., 209

Yamato Bank, 146
YASUI, M., 151
YODER, H. S., 149
YOKOYAMA, I., 151